STELLAR EVOLUTION, NUCLEAR ASTROPHYSICS, AND NUCLEOGENESIS

A. G. W. CAMERON

Edited by
David M. Kahl

Introduction by
Jordi José

DOVER PUBLICATIONS, INC.
Mineola, New York

Bibliographical Note

This Dover edition, first published in 2013, is an edited and newly reset reprint of the work originally published by Atomic Energy of Canada, Ltd., in 1957.

International Standard Book Number

ISBN-13: 978-0-486-49855-3
ISBN-10: 0-486-49855-7

Printed in Canada
49855701   2025
www.doverpublications.com

Introduction

The search for the energy source of stars, capable of maintaining long-lived luminosities, troubled renowned physicists during the 19th century. J. R. Mayer, J. J. Waterson, H. von Helmholtz, and W. Thomson (later known as Lord Kelvin) proposed that the energy radiated by the Sun comes from the conversion of gravitational potential energy into heat. The estimated lifetime of the Sun, from Kelvin's *meteor theory*, yielded about 30 Myr, an age clearly at odds with estimates based on geological records.

After the serendipitous discovery of radioactivity by A. H. Becquerel in 1896, the focus shifted towards nuclear energy.* In 1920, during a series of precise atomic mass measurements, F. W. Aston reported that four hydrogen nuclei are heavier than a helium nucleus. Shortly afterwards, A. S. Eddington (1920), on the basis of Aston's measurements, suggested that the energy source of the Sun was due to conversion of hydrogen into helium. The pioneering studies on Coulomb barrier penetration by G. Gamow (1928) and R. W. Gurney and E. U. Condon (1929), were a major step forward, and led R. Atkinson and F. Houtermans (1929) to conclude that quantum tunneling could allow energy generation in stars through nuclear fusion.

Work on hydrogen fusion reactions via *pp-chains*, by R. Atkinson (1936) and H. Bethe and C. L. Critchfield (1938), and through the *CNO cycle*, by C. F. von Weizsäcker (1938) and H. Bethe (1939), paved the road for the first self-consistent studies of element production in stars, the so-called *nucleosynthesis theory*, by F. Hoyle (1946, 1954). Equally influential was the compilation of Solar System abundances by H. Suess and H. Urey (1956): plotted as a function of mass number, A, the distribution of abundances revealed a complicated pattern, with hydrogen and helium being by far the most abundant species. The presence of several noticeable maxima in the region $A \geq 100$, for the distributions

---

*In 1919, H. N. Russell, using simple physical estimates inferred that the rate of the *unknown process* that supplies energy to stars must increase with increasing stellar temperature. He also concluded that this dependence of energy production on temperature would stabilize stars for long periods of time.

of both odd- and even-A species, was soon interpreted as due to nuclear physics effects (such as closed neutron shells in nuclei).

Still, observational evidence of stellar nucleosynthesis was missing ... The "smoking gun" was provided by the detection of technetium in the spectra of several S giant stars by P. W. Merrill (1952). Technetium has no stable isotopes. Since its longest-lived isotope has a rather short half-life ($T_{1/2}[^{98}\text{Tc}] \sim 4.2$ Myr) as compared with the Galactic timescale, the obvious conclusion was that it must have been produced *in situ*, in the observed stars, proving that nucleosynthesis is still ongoing in the Universe. Two seminal papers, that provided the theoretical framework for the origin of the *chemical* species, were published shortly after, almost exactly a century since Darwin's treatise on the origin of *biological* species: the first, by E. M. Burbidge, G. R. Burbidge, W. A. Fowler and F. Hoyle (known as *B²FH*), appeared in *Reviews of Modern Physics* in 1957; the second, a compilation of lecture notes known as the *Chalk River report CRL-41*, was a truly independent work by A. G. W. Cameron (1957).

Since those days, more than half a century ago, nuclear astrophysics has come of age as a truly multidisciplinary field, aimed at understanding energy production in stars and the origin of the chemical elements in the Universe. New tools and developments, at the crossroads of theoretical astrophysics, observational astronomy, cosmochemistry and nuclear physics, are continuously bursting our understanding in this area: supercomputers are providing astrophysicists with the required capabilities to study the evolution of stars in a multidimensional framework; the emergence of high-energy astrophysics with space-borne observatories has opened new windows to observe the Universe, from a novel panchromatic perspective; cosmochemists are continuously developing new techniques to isolate and analyze tiny pieces of stardust embedded in primitive meteorites, giving clues on the processes operating in stars as well as on the way matter condenses to form solids; nuclear physicists are determining reaction rates at (or close to) stellar energies, through combined efforts with stable and/or radioactive ion beams and theoretical modeling. Nevertheless, our modern conception of nuclear astrophysics owes credit and is deeply rooted in the two seminal papers by Cameron and B²FH.

Al Cameron was born on June 21, 1925 in Winnipeg, Canada. Trained in nuclear physics at the University of Saskatchewan, Merrill's discovery of technetium in giant stars shifted Cameron's interests into astrophysics, particularly in the aspects concerning energy production in stars and the origin of the chemical species. In Cameron's own words, Merrill's discovery raised the question "where did the stars get the neutrons?" In 1953, he already identified $^{13}\text{C}(\alpha, \text{n})^{16}\text{O}$ as the likely neutron source required for heavy element production in (red) giant stars. This was followed, shortly after, by the identification of a second neutron source, $^{22}\text{Ne}(\alpha, \text{n})^{25}\text{Mg}$, relevant for massive stars.

Indeed, his interest on the different neutron processes needed to shape the chemical abundance pattern in the mass range $A > 70$, the s- and r-processes (to which he originally referred to as *neutron captures on slow and fast time scales*), can already be found in his magisterial 1957 paper. Following the growing interest in space science research in the US, Cameron left Canada in 1961. He occupied different research and faculty positions in the US, including NASA's Goddard Institute for Space Studies, the Yeshiva University, the Harvard-Smithsonian Center for Astrophysics, and after "official retirement," the Lunar and Planetary Laboratory of the University of Arizona. His outstanding scientific contributions span a wide range of research interests in a field that he liked to refer to as *cosmogony*. This includes nuclear astrophysics (nucleosynthesis, mass formulae, nuclear level densities, thermonuclear and weak interaction rates, Solar System abundances, the r-process), stellar evolution (star formation, supernovae, neutron stars), and planetary science (the origin of the Solar System, the birth of the Moon-Earth system), among others.

While B^2FH is still available, the legendary Cameron's *Chalk River report CRL-41* has remained out of reach for many generations of nuclear astrophysicists. As a tribute but also as a fine example of a foundational paper, it is a great pleasure to see *Stellar Evolution, Nuclear Astrophysics, and Nucleogenesis* accessible once more to the increasingly large community of nuclear astrophysicists.

Al Cameron passed away on October 3, 2005 but his legacy will keep on inspiring and enlightening future generations of nuclear astrophysicists, or as he preferred to say, *cosmogonists*.

JORDI JOSÉ
Universitat Politècnica de Catalunya—BarcelonaTech,
and Institut d'Estudis Espacials de Catalunya

July 2012

Foreword

The enclosed is a newly edited and complete re-typesetting of the late Alastair G. W. Cameron's seminal work "Stellar Evolution, Nuclear Astrophysics, and Nucleogenesis," originally published by Atomic Energy of Canada Limited on behalf of Chalk River Laboratories in June 1957 and reprinted in December 1961, being referenced as either *AECL-454* or *CRL-41*.

Despite the author's modest description that the following are merely 'notes' for a series of lectures (given when he was 31 years old), the contents of this work are a major basis of what today is considered the field of stellar nucleosynthesis, published independently and concurrently to the innovative paper by Burbidge, Burbidge, Fowler, & Hoyle [1]. However, the present text has always been relatively less accessible than the latter, owing namely to its form of publication. Although Cameron also wrote a peer-reviewed and more available paper [2], it is a mere shade of what follows, being truncated about 90% in volume.

If *CRL-41* was not as readily available, nor consequently as often cited near its original date of publication, the passage of more than fifty years has only made the situation more dire. As the field of nuclear astrophysics continues to grow (particularly as it has in the past couple decades), it is paramount that the one of the essential and fundamental scientific works not continue along a path leading to obscurity. Upon finding an original, one may note that the document is not particularly reader-friendly, since it was set with a simple typewriter (with Greek letters and various mathematical symbols written in by hand). Despite the above and its dated nature, the document continues to offer important research, reference, and especially pedagogical value today, as well as into the foreseeable future.

It was with these reasons in mind that a project to ensure its availability in the present form was undertaken. The editor kindly acknowledges S. Aftergood, A. A. Chen, J. José and Y. K. Kwon for promoting this project, as well as stimulation from S. Kubono, and the productive environmet at the Center for Nuclear Study. As a testament to his legacy in the field of nuclear astrophysics, this republication is dedicated to the memory of Al Cameron.

DAVID M. KAHL

Wakō, Saitama, Japan

February 2012

Preface to the Second Edition

The writer has taken advantage of the issuance of a second edition of these notes to make corrections, both factual and typographical, in the text. There have also been some additional discoveries and calculations made in the last few months which have a considerable bearing on some parts of this material, and these are set forth in this updated edition. Material is inserted after the appropriate section of the text to which the supplementary note has principal reference. Supplementary references will also be found at the end, included with the other references.

The writer is indebted to the following people for discussions and correspondence in connection with material discussed in these supplementary notes: W. T. Sharp, G. E. Lee-Whiting, E. Vogt, A. J. Ferguson, H. E. Gove, A. E. Litherland, W. A. Fowler, H. D. Holmgren, R. Davis, R.L. Sears, J. L. Greenstein, R. A. Ferrell, R. L. Macklin, S. G. Thompson, W. P. Bidelman, and G. B. Herbig.

A. G. W. CAMERON
Chalk River, Ontario

January 1958

Preface to the First Edition

This report contains the notes for a series of lectures given in the Physics Department at Purdue University in the period March 25 to April 5, 1957. The writer is very grateful for the hospitality which he received at Purdue during these lectures.

Nuclear astrophysics is a very fast developing field. These notes have been prepared primarily for the use of present and potential research workers in this field. They attempt to present a review of the field with reasonable technical completeness and to indicate where further research is needed. While many new results are given here, this report is in no sense intended to be a publication of them, particularly since many of the results are very preliminary, and it is expected that they will be considerably improved in the near future. Nevertheless, several important conclusions about stellar evolution can be drawn from the results in preliminary form, and it has been felt useful to present conclusions at this time which will continue to be at least qualitatively correct.

No attempt has been made to polish the presentation of material in this report. The material has been basically designed for presentation in the form of lectures. Hence there has been no effort to document fully every statement, although a reasonable guide to the literature may be found in the references which are quoted.

The writer has benefitted greatly in his understanding of much of the material presented here from conversations and correspondence with many colleagues both at Chalk River and elsewhere. Particular thanks are due to J. L. Greenstein, M. Schwarzschild, W. A. Fowler, G.R. and E.M. Burbidge, L. G. Elliott, and W. T. Sharp.

A. G. W. CAMERON
Chalk River, Ontario

23 April, 1957

Table of Contents

Chapter 1

Introduction

In these lectures we will be interested in the development of different kinds of stars, in the nuclear reactions which can go on in their interiors, and in the bearing of these considerations on the chemical composition of the universe and the origin of the elements. We shall see that a good case can be made for the theory that the elements have been and are being made in stellar interiors. However, we must first briefly survey certain fields of astronomical knowledge which will particularly concern us.

1.1 Clusters of galaxies

The largest organized units of matter in space appear to be the clusters of galaxies. Most galaxies appear to belong to cluster. A cluster may contain only a few galaxies, but there is a continuous range of sizes extending up to the clusters with a membership of many thousand galaxies. The galaxies in the larger clusters have a greater internal velocity dispersion than do those in small clusters. The clusters themselves do not have large relative motions. Zwicky [3] has pointed out that there is a limit to the size of organized material units given by

$$D_{cl} < \frac{c}{v_g} D_g, \tag{1.1.1}$$

where D_{cl} is the diameter of a cluster, D_g is the average intergalactic distance in the cluster, v_g is the mean velocity of a galaxy, and c is the velocity of light. The largest clusters have diameters of the order of D_{cl}.

1.2 Galaxies

The most numerous class of galaxies is that of the elliptical galaxies. These appear to be composed mostly of very old stars with only relatively small amounts of interstellar gas and dust also present. The ellipticals vary in shape from those which are nearly spherical to those which are quite flattened. The spiral galaxies are more massive than the ellipticals. They have, in addition to a subsystem of old stars resembling an elliptical galaxy, spiral arms closely associated with large quantities of gas and dust. The spiral arms contain also many newly formed stars. Some smaller, irregular galaxies also exist; these contain mostly stars and interstellar material similar to that found in spiral arms. The mass of a typical galaxy is 10^{11} solar masses, but galactic masses range both up and down from this figure by large factors. The stars typically found in spiral arms are called Population I; those typically found in elliptical galaxies are called Population II stars.

1.3 Star clusters

There are major subsystems in galaxies called star clusters. These occur in two classes: the globular and galactic clusters. The globular clusters contain Population II stars. A typical cluster may contain 100,000 stars, but some of them contain many millions of stars. These cluster are nearly spherically distributed about the center of our galaxy. The galactic clusters are concentrated toward the plane containing the spiral arms, and they contain Population I stars. Globular clusters are so massive and so compact that they are very stable against disruption due to the gravitational effects of passing stars. On the other hand, galactic clusters can be dispersed in space by such encounters in the course of a few billion years. An extreme form of galactic cluster is the O or B Association, which consists of a few usually bright stars receding from each other with quite high velocities. These stars were in the neighborhood of a common point in space a few million years ago, but in a few more millions of years they will be so dispersed that it will be difficult to recognize them as an associated group of stars.

2

1.4 Stellar luminosities

There are many different classes of stars, and during the course of these lectures we shall try to reach an understanding of the paths of development of some of these stellar classes. One of the most remarkable properties of stars is the wide range of their luminosities. The brightest stars known emit about 10^6 as much energy per second as the sun; the faintest known emit only 10^{-6} as much energy per second as the sun. The total rate of energy emission of a star is called its absolute bolometric magnitude. Astronomical magnitudes form a logarithmic scale with a factor of 2.512 between magnitude classes. Because of the limited spectral transmission of the earth's atmosphere and the even more limited spectral sensitivity of most recording instruments, it is difficult to translate measurements into bolometric magnitudes. Hence in practice there are many systems of practical magnitudes which are based on specific spectral ranges. These magnitudes include the visual, photographic, and photoelectric magnitudes in various colors. The difference between the photographic and visual magnitudes of a star is called its color index. This gives a measure of the surface temperature of the star. The absolute magnitude of a star is equal by definition to the apparent magnitude it would have if it were placed 10 parsecs away from us. The parsec is a unit of astronomical distance equal to 3.258 light years.

1.5 Stellar spectra

The surface temperatures of the stars vary from slightly less than 2000 °K to more than 500,000 °K.* The elements present in the surface layers of a star occur in various stages of ionization at different temperatures; hence the appearance of a stellar spectrum is sensitive to the surface temperature of the star. The spectra of most stars can be arranged in the following continuous sequence (each class is subdivided into tenths):

Class O: Temperatures of 25,000 °K up. Lines of ionized helium.

Class B: 25,000 – 11,000 °K. The lines of hydrogen and neutral helium are conspicuous

---

*Editor's Note: The 13th Conférence Générale des Poids et Mesures decided that the unit of thermodynamic temperature is denoted as the unit *kelvin* and abbreviated by the symbol 'K' in 1967, prior to which time it was denoted *degree Kelvin* and with the symbol '°K' [4].

at B0. Ionized oxygen and ionized carbon become strong at B3. Neutral helium is strongest at B5. Hydrogen lines become progressively stronger in the higher numbered subdivisions of the class.

Class A: At A0, hydrogen and ionized magnesium lines are strongest while the helium and ionized oxygen lines have disappeared. Hydrogen lines weaken in the higher subdivisions, while ionized metals (Fe, Ti, Ca) strengthen. 10,700 – 7500 °K.

Class F: Class F0 is rich in lines of the ionized metals, the strongest being the H and K lines of singly ionized Ca. Metallic lines strengthen and hydrogen lines weaken as we pass through this class. 7500 – 6000 °K.

Class G: In this class the lines of the neutral metals become strong while the hydrogen lines continue to weaken. Molecular bands of CN and CH appear. 6000 – 4910 °K. Our sun is class G2.

Class K: In general, molecular bands and lines of neutral metals become much stronger while the lines of hydrogen and ionized metals continue to weaken. At K5 the lines of TiO are weakly visible. 4910 – 3500 °K.

Class M: The characteristic feature is the complex spectrum of molecular oxide bands, of which the TiO bands are the strongest. 3500 – 2200 °K.

Some stars do not fit into this sequence of spectral classes. Therefore, some additional spectral classes have been established which parallel the previous classes in temperature. These classes include the following:

Class S: A low temperature class parallel to the Class M. This is still characterized by molecular oxide bands, but the most prominent feature is the ZrO bands. Certain elements such as Zr, Y, Ba, La and Sr give strong atomic lines and oxide bands. Lines of neutral technetium are usually seen.

Classes R and N (or Class C): Parallel in temperature to the ordinary classes K and M. The spectrum is characterized not by oxide but by molecular carbide bands, such as those of CN, C_2, and CH.

4

Class W: Extremely high temperature objects called Wolf-Rayet stars with bright, broad, hazy emission lines of ionized helium and highly ionized carbon, oxygen, and nitrogen. Two sequences exist: the WC stars have strong carbon lines and weak nitrogen lines; in the WN class the reverse is true.

1.6   Surface temperature-luminosity diagrams

It has been known for a long time that if one plots some form of stellar magnitude against some measure of surface temperature, then most points tend to cluster along certain lines in the diagram. This type of diagram is often called a color-magnitude diagram or a Hertzsprung-Russell (H-R) diagram. In the last few years it has been found that characteristically different H-R diagrams are obtained for globular and galactic clusters and for other types of associated objects. These will be discussed in greater detail in Chapter 3. However, at this point it may be stated that most stars cluster about a single line on these diagrams, called the main sequence.

1.7   Mass-luminosity relation

If two stars are joined to form a binary pair and if the periods, dimensions, and orientations of their orbits can be determined, then the masses of the stars follow from Kepler's third law. Quite a few stellar masses have been determined in this way for visual binary pairs. Russell & Moore [5] have given us an empirical relationship between the mass and luminosity of main sequence stars. If the intrinsic luminosity L and the mass M are measured in solar units, then

$$\log L = 3.82 \log M - 0.24. \tag{1.7.1}$$

The constant -0.24 expresses the fact that the sun is overluminous for this relationship by 0.60 astronomical magnitudes.

1.8 Stellar composition

There appears to be a striking parallelism between the composition of the earth and the meteorites and the sun. This parallelism will be discussed in much greater detail in Chapter 9. However, at this point we may note that the earth and meteorites have lost volatile substance consisting of gases and substances with low boiling points, such as mercury. This actually accounts for most of the mass because it is found that the sun is nearly all hydrogen with about five percent of helium by numbers of atoms and much less than one percent of other elements. Suppose we denote the atomic ratio of the metals to hydrogen in the sun by $R_{\odot}$. One of the most important discoveries in modern astrophysics is that this ratio is not a constant of matter in cosmic proportions. Stars which can be classed as extreme Population II commonly have ratios of about $R_{\odot}/10$ or $R_{\odot}/20$. In some extreme objects the ratio appears to be still much smaller. On the other hand, in stars which can be classed as extreme Population I, the ratio is larger than in the sun with values of $2R_{\odot}$ to $4R_{\odot}$ being common. The light elements such as C, N, O and also heavier elements, such as Ba, appear to vary in abundance by about the same factors as the metals.

1.8.1 Supplementary Notes: Stellar populations

A conference on the problem of stellar populations was held at the Vatican in the spring of 1957. The general consensus of opinion at this conference appears to have been that at least five distinct stellar populations must now be recognized (*cf.* the summary paper by F. Hoyle in the proceedings of this conference, now in press[†]). The oldest stars in the galaxy, with ages of about 6 to 7×10^9 years, include two populations, called Halo Population II and Intermediate Population II (some people prefer to call the former Extreme Population II). The Halo group is nearly spherical, extends very far out into space, and contains a very variable content of elements heavier than helium averaging about 0.3 percent by weight. The Intermediate Population II stars are a somewhat more flattened group with heavy element contents of the order of one percent or a little less. These populations contain about 15 to 20 percent of the mass of the galaxy.

[†]Editor's Note: Published as [6].

Most of the mass of the galaxy lies in the Disk Population, a very flattened group of stars with ages 4 to 6×10^9 years. This group includes our sun and has a heavy element content of one to two percent.

The younger stars belong to the Older Population I and the Extreme Population I (some people prefer to call the former the Intermediate Population I). Perhaps about 10 percent or a little less of the mass of the galaxy consists of these populations of stars. The heavy element content ranges up to about four percent. These stars are closely associated with the gas and dust in the spiral arms of the galaxy.

Only about two percent of the mass of the galaxy is in the form of gas and dust. This small amount of material is more nearly consistent with the amount of heavy element enrichment which can have occurred from what is known about the rate of star births and deaths. Even so, it appears likely that the frequency of supernova explosions must have been considerably greater in the early history of the galaxy than it is at present.

1.9   Anomalous stellar abundances

Many different stars have individual abundance anomalies which differ markedly from the solar abundances or the systematic variations of solar abundances characteristic of extreme Populations I and II stars. The stars of spectral class S appear to have unusually large abundances of most elements heavier than about germanium. These heavy elements are also overabundant in a small group of stars called BaII stars (the designation "II" means that the barium is singly ionized; "I" would have indicated neutral barium). Carbon is very overabundant in stars of spectral classes R and N in "CH" stars and to a lesser extent in BaII stars and in class S. The "peculiar A" stars, or spectrum variables, have strong surface magnetic fields and an associated large overabundance of heavy elements and, in particular, of the rare earths. Wolf-Rayet stars (class W) appear to be deficient in hydrogen but rich in helium, carbon or nitrogen, oxygen, and, often, neon. We shall be interested in seeing how these abundance anomalies can arise as a result of nuclear reactions which go on in stellar interiors and at stellar surfaces.

1.10   Interstellar matter

Something like half of the mass of our galaxy exists in the form of gas and dust between the stars. Most of this gas is in the form of atomic hydrogen. To the extent that the abundances of other elements can be determined, they are of the same order of magnitude as in the sun. The interstellar matter is a very chemically reactive medium, and much of the heavier elements have formed chemical compounds which have collected to form dust grains. There is a considerable uncertainty as to the size of these grains. The interstellar absorption of starlight is caused by the scattering properties of the dust grains.

1.11   The ages of the sun and stars

The presence of naturally radioactive material on the earth shows that the material from which it was formed does not have an infinite age, but that there was a process in which the elements were formed. Considerations of the amount of energy which can be released in nuclear reactions allows limits to be placed on the length of time that stars can radiate energy at their presently observed rates. There are many dynamical properties of the stars in the galaxy which are functions of a galactic time scale. These include the time required to disrupt galactic clusters and binary stars. From these and many other considerations, it appears that the Population II stars in our galaxy were formed about 7×10^9 years ago. The Population I stars have been forming continuously since then. Our sun is an old Population I star. Our planetary system has an age of 4.5×10^9 years, and another period of at least 0.5×10^9 years passed between the time of formation of the elements from which the solar system was formed and the time at which the formation process caused chemical separations of uranium and lead. The sun probably formed at the same time as the planetary system. The dynamical properties of O Associations show that their stars have been formed as recently as about 10^6 years ago.

8

1.12   Ejection of mass from stars

We have seen that the stars are being continuously formed out of the interstellar medium. They are also in the process of ejecting material to the interstellar medium. There are three principal mechanisms by which this can take place:

Supernova explosions: At the peak of its light curve, a supernova usually outshines the galaxy in which is it situated. Spectroscopic observations show that a substantial fraction of a solar mass is ejected in a supernova explosion. The ejected gases have outward velocities of some thousands of kilometers per second. About one supernova explosions per 300 years occurs in an average galaxy.

Nova explosions: These are much less spectacular than supernova explosions; only about 10^{-4} as much light is emitted. About 0.001 solar mass is ejected in an outburst. Some of the novæ which had smaller light amplitudes have been observed to flare up more than once, and it is likely that most novæ are recurrent. The time between recurrences is of the order of a century, but there is a very large dispersion in this interval.

Continuous ejection: After a star has evolved away from the main sequence, there appear to be many stages in which clouds of gas an be emitted from the surface. Deutsch [7] has observed a case in which a class M supergiant star is emitting matter at the rate of one solar mass per 30,000,000 years or less. Even white dwarf stars have been observed to be associated with intense magnetic fields on the surfaces of the emitting objects.

1.13   The formation of the elements

We have seen that stars are continually being born in interstellar space and that in the later stages of their evolution they return matter to the interstellar medium. Salpeter [8] has estimated that the amount of matter which has been formed into stars and has been returned to the interstellar medium is of the same order of magnitude as the total mass of all presently existing stars. We have also seen that the newly-formed stars have much larger abundances of heavier elements than the old stars of Population II, and hence, presumably, the interstellar medium has been similarly enriched in heavier elements. During the course of these lectures we shall see that there are many types of nuclear reactions which can go on in stellar interiors that can produce certain species of the elements with their observed relative abundances.

It is a natural conjecture from the above considerations that the elements have been formed from hydrogen during the life of the galaxy. Thus about 7×10^9 years ago it is suggested that the galaxy was a large mass of gas (all or mostly hydrogen) in space. Much of this gas condensed to form Population II stars; the first stars to condense were very massive and evolved very quickly, ejecting heavier elements into the interstellar medium. These elements were incorporated into other stars and further processed by nuclear reactions. The process of element formation should therefore still be going on in stars today.

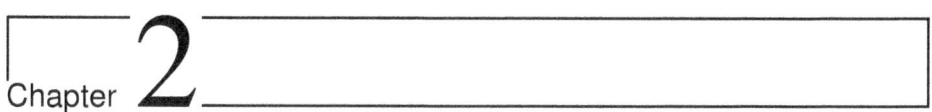

Chapter 2

The Development of the Galaxies

Strictly speaking, the theories of element formation to be described are nearly independent of cosmological considerations because element formation is believed to take place after the formation of stars in the galaxy. This is compatible with the leading cosmological theories in the present stage of the development of the theories of the origin of the elements. However, there are some cosmology-linked effects that are susceptible to both observations and calculation, and it seems useful to review briefly some of the ideas about cosmology and the development of galaxies.

2.1 Cosmology

To illustrate the main difference likely to be important to theories of element formation, we can briefly consider the cosmological exploding models (without a cosmological constant) of Friedman [9], and the steady-state model of Bondi & Gold [10] and Hoyle [11; 12]. These models give different second-order terms in the expected red-shift-magnitude relation for distant galaxies. Present techniques of photoelectric photometry are promising to subject these second-order differences to observational test.

In the exploding models, the universe was once in a highly compressed state. It was at one time thought [13] that the conditions existing in the compressed state were suitable for formation of the elements in a non-equilibrium process of neutron capture. The difficulty with such theories is that there is no means of building nuclei heavier than He^4 unless one goes to unrealistically high densities of matter. Nevertheless, it may be that one should expect primordial abundances of deuterium, He^3, and He^4 to be formed in the early stages of an exploding universe. This can be tested in principle by measuring the helium abundances and deuterium abundances in Population II stars. The problem is complicated because we know that helium is one of the main products formed in stellar thermonuclear reactions, and helium abundances are very difficult to determine accurately from spectroscopic measurements. Deuterium is even harder to measure in the stars because its lines are shifted only slightly from those of hydrogen, and its abundance is small compared to that of hydrogen. Perhaps we will have to wait for radio astronomy techniques to improve in sensitivity by another order of magnitude so that one can hope to detect the neutral deuterium in interstellar space.

In the early stages of the exploding universe, the density of radiation exceeds that of matter. Gamow [14] has pointed out that no local condensations of matter are possible until the expansion has proceeded far enough for the radiation density to fall below the matter density. Then local condensations are possible if the matter density is large enough.

In the steady-state model, the average density of matter in space is constant in time, the expansion of the galaxies being balanced by the creation of new matter in space. In this picture, the new gas formed in intergalactic space, augmented by matter ejected from stars in small galaxies, condenses to form new galaxies. The gas in these new galaxies would have small abundances of all the elements present initially. These small abundances would have important effects in the stars forming in the new galaxy which may be tested by calculations in several ways.

Preliminary photoelectric red-shift measurements of Baum [15] show that the red-shift-magnitude relationship is very nearly linear out velocities of recession of $0.4c$. These measurements support one of Friedman's models in which the expanding universe is closed; it is slowing in its expansion and will eventually stop and then start contracting. The data seem to be nearly good enough to disprove the steady-state model.

2.2 Fragmentation of gas clouds into galaxies and stars

We have seen that either of the cosmological theories is likely to produce a mass of gas, mostly hydrogen, which is unstable to a contraction into galaxies. Hoyle [16] has given an attractive theory for the course of development of this gas. His theory is outlined presently in brief detail.

The key to Hoyle's galactic evolutionary theory lies in the thermal properties of the hydrogen gas. Let us consider a hydrogen gas cloud with an initial density $\rho_0 = 10^{-27}$ g/cm^3. To estimate the temperature of the gas, we must realize that most cosmic objects have very large Reynold's numbers, and hence the gas is likely to be turbulent, i.e. to develop mass motions. The energy of the mass motions becomes converted into heat. Hoyle estimates that certainly more than 10^{12} ergs/g, and possibly more than 10^{14} ergs/g, is made available in this way.

The thermal energy of atomic hydrogen is $\frac{3}{2}RT$ per gram, where R is the gas constant and T is the temperature ($^\circ$K). Hence somewhat more than 10^{12} ergs/g is required to heat the hydrogen from 0 $^\circ$K to 10,000 $^\circ$K. At this point, ionization of the hydrogen sets in. Ionization is nearly complete at 25,000 $^\circ$K. It requires somewhat more than 10^{13} ergs/g to ionize hydrogen.

Hydrogen gas can radiate energy in three ways: (1) line emission, which will be neglected; (2) free-free transitions, in which unbound electrons make transitions between two states of the continuum; and (3) free-bound transitions, in which electrons are captured from a state of the continuum into a bound state of the hydrogen atom. The rate of hydrogen radiation from the latter two processes is

$$1.45 \times 10^{-27} \, T^{1/2} \left[1 + \frac{3.85 \times 10^5}{T} \right] N^2 \text{ ergs/cm}^3 \text{ s,}$$

where N is the number of hydrogen atoms per cm^3, and T is the kinetic temperature in $^\circ$K. This formula holds well only for $T > 25,000$ $^\circ$K. We see that the rate of radiation decreases until very large temperatures are reached.

From these properties of hydrogen gas, we can see that the large mass of hydrogen gas with which we have to deal is likely to have a temperature of either close to 10,000 $^\circ$K or greater than 10^5 $^\circ$K. The reason for these two temperature classes is that the time of the development of a galaxy is of the order of 10^{17} seconds, and in this time hydrogen, initially at temperatures $< 10^5$ °K, can cool by radiation until most of the hydrogen atoms have recombined with electrons.

Let us consider first the case of a gas cloud at a temperature of 10,000 °K. The necessary condition for a gas cloud of uniform composition to be in thermal equilibrium, in the case where radiation pressure is negligible so that gravitational forces are balanced by gradients in the gas pressure, is that the total gravitational potential energy Ω (taken as positive) must be equal to twice the thermal energy of the gas. If Ω is greater than this, then a spherical cloud will contract. This gives us the following contraction condition:

$$\frac{GM}{V^{1/3}} > 3RT, \tag{2.2.1}$$

where M is the total mass, V is the total volume of the gas cloud, and G is the universal gravitational constant. This inequality gives a lower limit to the mass of a gas of neutral hydrogen which can condense. For $M = \rho_0 V = 10^{-27}V$, then (2.2.1) gives $M > 3.6 \times 10^9$ solar masses as the minimum mass of hydrogen gas which can condense. During such condensation the thermal energy is quickly radiated away, and the temperature remains at 10,000 °K. Hence the contraction of the gas takes place under isothermal conditions.

The thermal energy released in a uniform isothermal contraction from a volume V to a volume V' is

$$MRT \ \ln \frac{V}{V'}. \tag{2.2.2}$$

The gravitation energy released by the contraction is of the order of

$$\frac{GM^2}{V^{1/3}} \left[\left(\frac{V}{V'} \right)^{1/3} - 1 \right]. \tag{2.2.3}$$

Now, if the initial mass of the hydrogen gas cloud is large compared to the minimum of (2.2.1), then it follows that (2.2.3) is large compared to (2.2.2), regardless of the value of V'. Hence in the contraction of such a cloud, only a small part of the gravitational potential energy is converted into thermal energy, and the rest must go into mass motions, which will be adequate to expand the cloud back to nearly its original dimensions. Such a cloud, therefore, cannot undergo appreciable contraction. However, such a cloud is also

14

unstable to the fragmentation into subcondensations which have about the minimum mass consistent with (2.2.1). We may identify such a fragmentation process with the formation of a cluster of galaxies with the subcondensations having a mass $\sim 3.6 \times 10^9$ solar masses under the previous assumptions. These are about the masses of elliptical galaxies.

Let us now consider the condensation of a galaxy of the minimum mass. Expressions (2.2.2) and (2.2.3) remain comparable for $V' = 0.1V$; the former is still three percent of the latter for $V' = 10^{-3}V$, and it has fallen to five percent of the latter for $V' = 10^{-6}V$. Hence there can be an almost complete dissipation of the energy of contraction for a decrease of linear dimensions by a factor of two or three, but further contraction is then prevented by the development of mass motions. However, after a contraction of linear dimensions by about a factor of three, the galaxy is then unstable to a fragmentation into further subcondensations. These, in turn, can contract in linear dimensions by about a factor of three, followed by further fragmentations, and so on. As the density of the subcondensations increases, the rates of contraction and radiation also increase. If a subcondensation contracts by a factor of $k^{2/3}$, and then divides into k equal masses having a radius equal to $1/k$ of that of the original subcondensation, and if many such stages of condensation take place, then Hoyle points out that the time for all stages of condensation is, in units of that required for the first stage,

$$1 + \frac{1}{k} + \frac{1}{k^2} + \frac{1}{k^3} + \ldots = \frac{k}{k-1}.$$

For $k^{2/3} \approx 3$, $k \approx 5$, and the time required for an infinity of steps is only 25 percent longer than that required for the first step. Hoyle postulates that the condensation of the first step will take a time of the order of 10^{17} seconds, and, since the fragmentation will really be a continuous process, very small subunits will be formed over the whole of this time scale.

The process of fragmentation will stop when the density of the subunits has become so large that they are opaque to their own radiation. Hoyle shows that such final subunits have masses which generally should lie in the range 0.3 to 1.5 solar masses with some dispersion which will extend outside this mass range. After a galaxy has been fragmented into a hierarchical structure of subcondensations, these must fall towards the galactic center of gravity where the hierarchical structure will be broken up by mutual interactions. This causes a shrinkage of the dimensions of the galaxies. Occasionally, certain subcondensations will resist disruption, and it is tempting to identify these with the globular clusters.

Let us now consider a gas cloud which has a temperature initially in excess of 10^5 °K. The minimum mass required for condensation of this gas cloud is given by

$$\frac{GM}{V^{1/3}} > 6RT, \qquad (2.2.4)$$

where the altered form of the heat content refers to fully ionized hydrogen. For $M = 10^{-27} V$ the minimum mass is 5×10^{11} solar masses. As contraction occurs, the rate of radiation starts to exceed the rate of release of thermal energy, and, eventually, a point of temperature instability must be reached that swiftly reduces the temperature to 10,000 °K. Assuming that the linear dimensions have decreased by a factor of five before this occurs, then the first fragmentation of the gas cloud will be into dwarf galaxies with masses of about 3×10^8 solar masses. These dwarf galaxies will collide with one another and most will be disrupted; some will escape into intergalactic space. We may identify the massive spiral galaxies with a formation process of this type.

Not all the gas in these galaxies will be formed into stars; there will be less tendency for condensation in regions of smaller gas density. Such gas will contract to the center of the galaxy, or, if the galaxy has a fairly large amount of total angular momentum, then the residual gas will contract to a disk passing through the center of the galaxy. Further star formation can then take place.

Chapter 3

Hertzsprung-Russell Diagrams

It has been remarked in Section 1.6 that the stars are not uniformly spaced in a diagram in which their luminosity is plotted against their surface temperature. Many conclusions about the course of stellar evolution can be drawn from a study of these Hertzsprung-Russell diagrams for individual star clusters. The methods of plotting the diagrams take a wide variety of forms: the luminosity can be expressed in bolometric, photographic, visual, or photoelectric magnitudes in different colors. The surface temperature can be expressed as a temperature, spectral class, color index, or as a difference of photoelectric magnitudes in different colors. However, the basic star sequences in these different types of H-R diagrams are very similar with only relatively minor distortions being introduced by the different forms of presentation.

3.1   H-R diagrams for galactic clusters

Figure 3.1.1 shows a composite diagram of several galactic clusters [17]. The common line which these cluster share on the left is the main sequence. The short segments of these diagrams on the upper right lie in the red giant and red supergiant region. It may be noted that all the galactic clusters except M67 have a gap between the sequences of stars lying on or near the main sequence and their red giant branches. This is called the Hertzsprung gap. The gap is wedge-shaped, being non-existent for M67 and becoming larger as one goes upwards in the diagram.

17

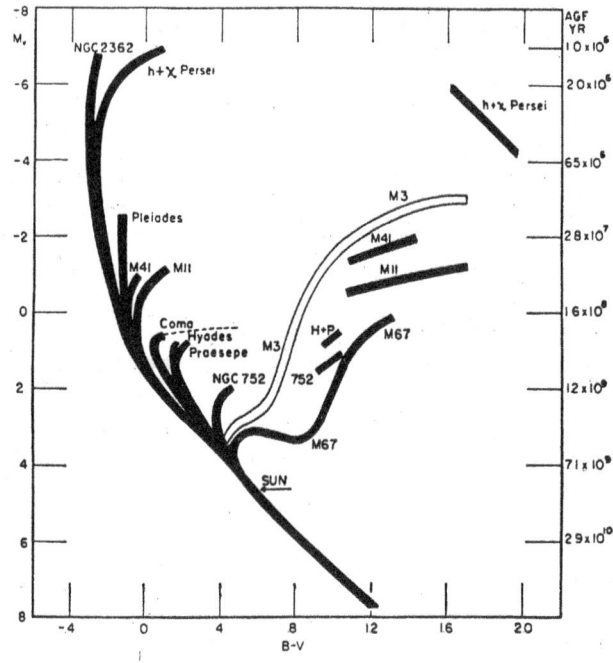

Figure 3.1.1: A composite color-magnitude diagram for galactic clusters. In this diagram visual magnitudes are plotted against the difference of the photoelectric magnitudes measured in blue and visual light. For comparison, the giant branch of the globular cluster M3 is also included. The lengths of time required for stars to evolve away from the main sequence are shown on the right.

Many features of stellar evolution may be deduced from an inspection of Figure 3.1.1. The stars at the top of the diagram are using up their energy sources most rapidly. Hence the clusters at the top of the diagram must have been formed very recently. We can see that the brighter stars in these clusters lie on sequences which curve away from the main sequence upwards and to the right. This establishes the direction of the evolution of the individual stars when they evolve away from the main sequence. After such stars reach the boundary of the Hertzsprung gap, they must evolve to the right very quickly to reach the giant branch appropriate to the cluster. They then sit on the giant branch for a considerable period of time. It may be noted that, in most of the cluster, the giant stars have about the same luminosity as those which are evolving off the main sequence. However, in an old galactic cluster like M67, the stars brighten by about three magnitudes

18

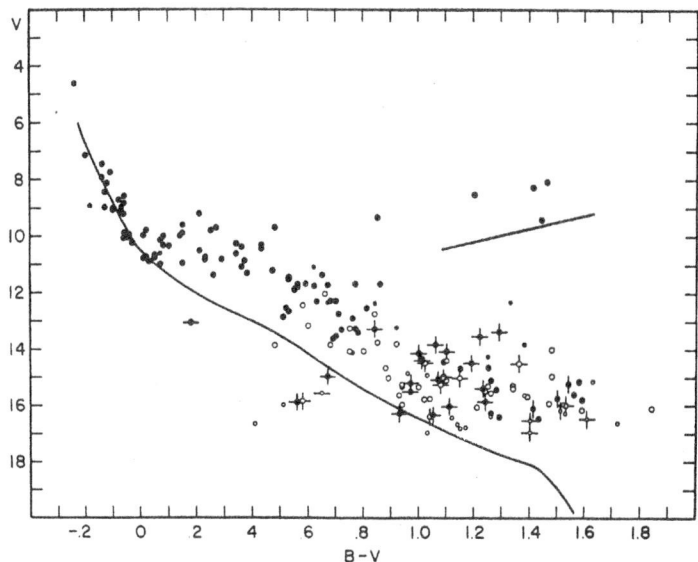

Figure 3.1.2: Color-magnitude diagram of NGC 2264. Here photoelectric visual magnitudes are plotted against the difference of blue and visual magnitudes. The stars are reddened by interstellar absorption. Dots represent photoelectric and circles represent photographic observations. Vertical bars show known light-variables and horizontal bars show stars with bright emission lines. The lines show standard main sequence and giant branches for Population I stars.

after evolving away from the main sequence. The lengths of time one can expect the stars to remain on the main sequence are shown at the right of Figure 3.1.1. The stars in galactic clusters belong to Population I, while those in clusters, such as NGC 2362 and h + χ Persei, have been formed very recently and are extreme Population I.

Our composite diagram does not suffice to tell us anything about the tracks of the stars before they reach the main sequence or after they have exhausted their sources of energy on the giant branch. Since we expect that the stars form from gas clouds contracting from interstellar space, then, initially, such contracting gas clouds must have very low surface temperatures owing to the large surface area available to radiate their energy released in the contraction. This would place the forming objects far to the right of the main sequence in the H-R diagram. As the contraction proceeds, the stars move towards the left until they reach the main sequence. Since the more massive stars radiate energy enormously faster than the less massive ones, we can expect them to reach the main sequence first. This raises the possibility that one might be able to find galactic clusters so recently formed that their brightest members would lie on the main sequence, and the faintest members would like to the right of the main sequence. Walker [18; 19] has found sever such clusters. The color-magnitude diagram found by him for NGC 2264 is shown in Figure 3.1.2. The brightest stars lie on the main sequence; it may be noted that the five giant stars here are less luminous than the brightest main sequence stars. The fainters stars lie above and to the right of the main sequence. Many of these stars show light variability and bright hydrogen emission lines. This places them in the class of "T-Tauri variables," which are believed to be stars in gravitational contraction which are still interacting with the interstellar medium from which they were formed.

3.2 H-R diagrams for globular clusters

There are many characteristic individual differences in the color-magnitude diagrams for the globular clusters. These clusters contain Population II stars formed a long time ago; like the galactic cluster M67 they do not show a Hertzsprung gap. M67 is compared with the globular clusters M3 and M92 in Figure 3.2.1. The main sequences of the globular clusters are not shown, but it is known that they lie to the left and below the main sequence of the galactic clusters (it appears that the main sequences of different globular clusters lie in somewhat different positions in the H-R diagram). Stars evolve away from the main sequence at about the same luminosity as in M67, but they increase in luminosity to a much greater extent when they reach the red giant region. It can be seen from Figure 3.2.1 that after reaching the tip of the giant branch at the upper right, the stars reverse their direction of evolution, and decrease somewhat in luminosity, and move to the left in the diagram. This region of the diagram is called the horizontal branch. The horizontal branch contains a region, shown by the gap in the M3 diagram, where the stars are light variables; such stars are called cluster variables.

Baum [20] points out that the features of the color-magnitude diagram that differ from one globular cluster to another include (1) the position of the main sequence; (2) the color of the top of the red branch; (3) the shape and tilt of the red giant branch; (4) the number of cluster type variables; (5) the populousness of the horizontal branch at the

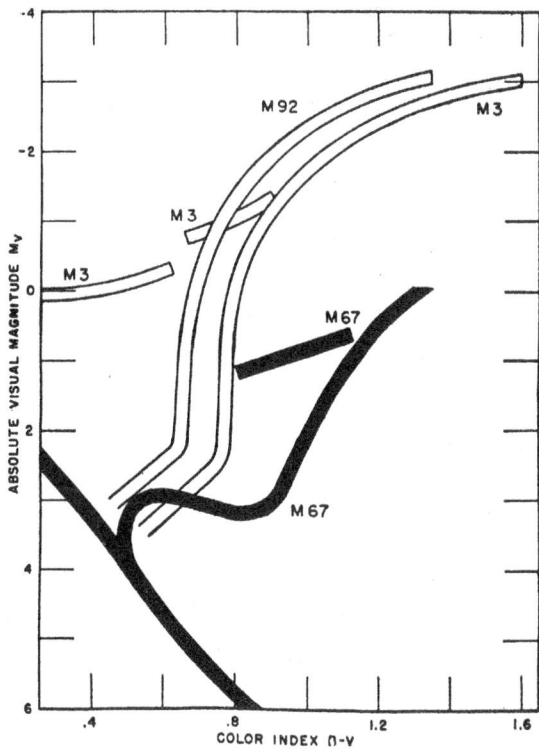

Figure 3.2.1: Color-magnitude diagrams of the galactic cluster M67 compared with the globular clusters M3 and M92. The absolute visual magnitudes are plotted against the difference of the photoelectric blue and visual magnitudes.

right of the cluster-type variables; and (6) the shape and tilt of the horizontal branch. These differences appear to be connected with the differences in chemical composition between the stars in different globular clusters, and there are corresponding correlations in the differences in these features in different clusters. The abundance ratio of metals to hydrogen varies quite a bit between different globular clusters but is always much less than in the sun.

Greenstein [21] has made extensive studies of the spectra of subdwarf and white dwarf stars. The subdwarf stars are members of extreme Population II which lie below the Population I main sequence. The H-R diagram for these stars is show in Figure 3.2.2. It may be seen that the stars with the smallest abundance ratios of metals to hydrogen

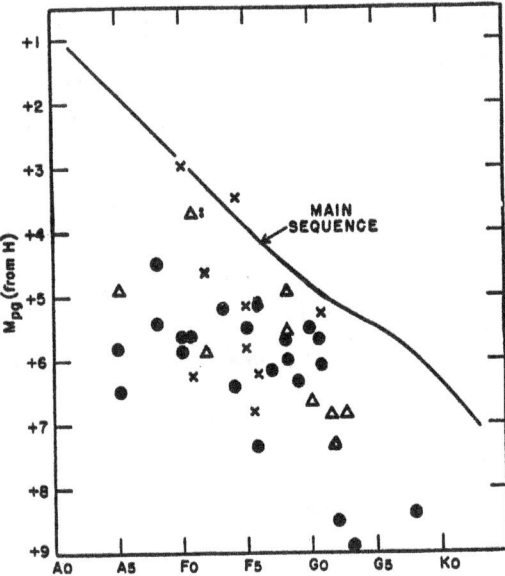

Figure 3.2.2: Hertzsprung-Russell diagram of the subdwarf stars. The photographic magnitude is plotted against the spectral class. The solid dots show stars with a smaller abundance ratio of metals to hydrogen than the open triangles. The line shows the position of the main sequence for Population I stars.

generally lie farthest below the Population I main sequence.

3.3 Evolutionary tracks in the H-R diagram

From an inspection of the Hertzsprung-Russell diagrams for globular and galactic clusters, we have reached certain conclusions about the evolutionary tracks of the stars in those diagrams. Sandage [22] has obtained evolutionary tracks for globular cluster stars by mapping stars from their original positions on the main sequence to their presently observed positions. These evolutionary tracks are show in Figure 3.3.1. They confirm the previous discussion. It should be noted that the stars in the giant branch have brightened so much that their evolution is very rapid. Such stars have nearly the same mass.

After stars travel to the left on the horizontal branch they are often quite unstable, shed mass, and show peculiar abundances of certain elements. Pre- and post-novæ exist

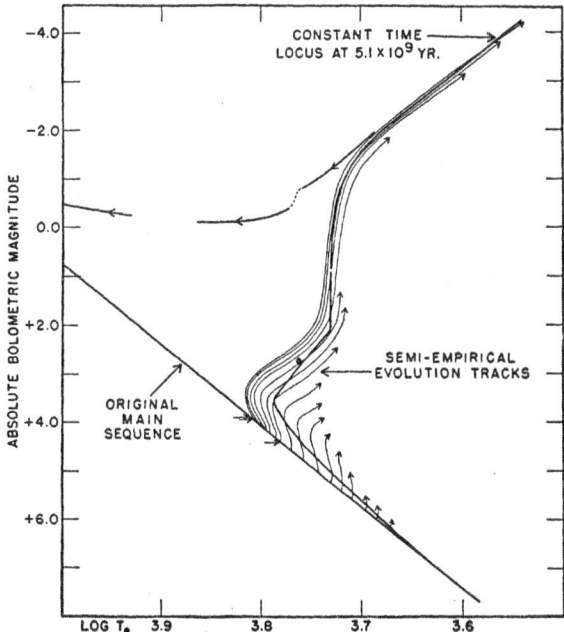

Figure 3.3.1: Evolutionary tracks in the Hertzsprung-Russell diagram for globular clusters. Absolute bolometric magnitudes are plotted against stellar surface temperatures. The heavy solid lines show the positions of the original main sequence and also of the present locus of stars in the globular cluster M3.

in the far left region of the H-R diagram; the nova explosions appear to be connected with internal instabilities which arise near the end of their course of evolution. It appears that the stars finally take the form of white dwarfs at the end of their course of evolution. Such stars are composed of a degenerate electron gas. They have a very high density and contain no hydrogen or energy sources in the interior. Their internal heat energy is gradually radiated into space and they are expected to cool off at a constant radius. Luyten [23] has plotted a color-magnitude diagram of the white dwarf stars; this is shown in Figure 3.3.2. The white dwarf stars have a very low luminosity, but their small diameters imply a small surface area, and hence they are usually blue objects. It may be seen that most white dwarfs appear to have diameters varying in size between one and four times that of the earth. The white dwarf stars are very numerous in space, second only to the numbers of stars on the main sequences. They represent the graveyard of

23

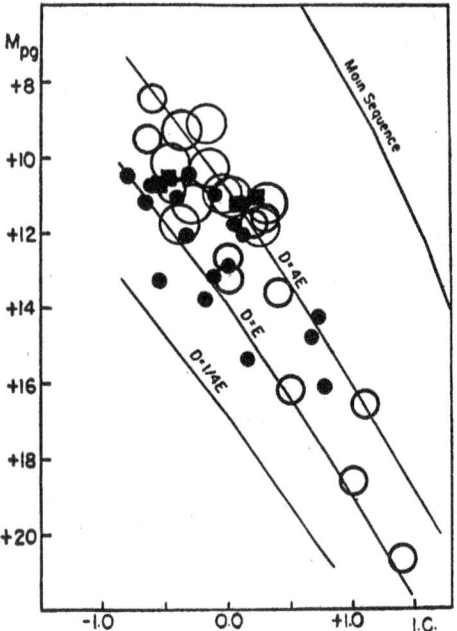

Figure 3.3.2: Color-magnitude diagram of the white dwarf stars. The photographic magnitudes are plotted against the color index. The thick line on the upper right shows the Populations I main sequence. The other lines give the diameter of the white dwarf in units of that of the earth. The track of a star cooling at constant radius would lie parallel to these lines.

stars which have evolved off the main sequence and exhausted all their energy sources.

24

Chapter 4

Physical Conditions in Stellar Interiors

Efforts have been made for many years to determine the physical conditions at various points in stellar interiors. These attempts have improved with the passage of time, and the more realistic inclusions of various physical effects in the theory. The development of the subject has been governed by the fact that the differential equations describing the stellar interior are nonlinear. Thus, some elegant mathematical analyses have been developed which enable stellar models to be computed using analytic approximations to certain physical properties; these models have the nice property that computations for one assumption of the mass and luminosity can be transformed to other values of mass and luminosity by using certain dimensionless variables in the calculations. These methods have given very useful results, but they will not be described here because stellar models are now being calculated numerically on electronic computers, and different methods of analysis are best suited to computer calculations.

4.1 Differential equations of the stellar interior

In most main sequence stars, atoms in the interior are largely stripped of electrons, and hence the gas obeys perfect gas laws except under conditions of very great density ($> 10^4$ g/cm^3) where electron degeneracy sets in. In order to write down the differential equations for the stellar interior, let us consider a thin shell of radius r about the center of the star with thickness dr. Then we get a condition for hydrostatic equilibrium in this layer by stating that the difference of pressure between the top and bottom of the layer must support the weight of the layer:

$$\frac{dP}{dr} = \frac{GM(r)}{r^2}\rho,$$

where $P =$ total pressure (gas pressure p_g + radiation pressure p_r),

$\rho =$ density in layer

$M(r) =$ mass inside layer,

and $G =$ universal constant of gravitation.

The conservation of mass in the star is given by the relation:

$$\frac{dM(r)}{dr} = 4\pi r^2 \rho.$$

The conservation of energy in the star is given by the relation:

$$\frac{dL(r)}{dr} = 4\pi r^2 \rho E,$$

where $L(r)$ is net outward flux of energy entering the bottom of the layer, and E is the energy production per gram per second in the layer.

We will obtain another differential equation which describes the transport of energy in the stellar interior. In general, there are three methods by which energy transport can take place: conduction, radiation, or convection. Conduction is not important unless electron degeneracy is present, a case that will be considered later. Let us consider the case where energy transport is solely by radiation.

Energy transport by radiation takes place more rapidly for large temperature gradients than for small. It is impeded by high densities and high opacities of the material through which it must pass. These conditions are expressed in an equation for the

26

temperature gradient required to transport a stated flux of energy [24]:

$$\frac{dT}{dr} = -\frac{3}{4ac}\frac{K\rho}{T^3}\frac{L(r)}{4\pi r^2},$$

where $T =$ the temperature ($^\circ$K),

$a =$ Stefan's radiation constant,

$c =$ the velocity of light,

and $K =$ the opacity of the material, which is an appropriate mean of the mass absorption coefficients of the materials present.

This is discussed in detail presently.

Now let us consider when energy transport by convection will take place. Let us consider a volume element of the gas which we will take large enough so that adiabatic conditions will hold if we give this element a sudden upward displacement. In this displacement, pressure equilibrium will be maintained. Under adiabatic conditions

$$\rho \propto P^{1/\gamma},$$

where $\gamma = \frac{C_p}{C_v}$ is the ratio of specific heats.

If the new density is less than that of the surrounding matter, the element will continue to rise, and hence that region of the star is unstable to the onset of convection. The condition for stability against convection is that the actual pressure gradient in the star must be less than that corresponding to a density gradient in adiabatic equilibrium. In general, such stability requires relatively small temperature gradients.

Under adiabatic conditions, the temperature varies with the pressure according to the relation

$$T \propto P^{(\gamma-1)/\gamma}.$$

A rising volume element has a smaller density than its surroundings, and hence it has a temperature excess. At higher levels it loses its excess heat by viscous effects and by radiation. Similarly, a descending element has too low a temperature and is heated in lower levels. This energy transport process will very quickly reduce a very steep temperature gradient until it is close to the limiting adiabatic gradient. Hence it may be seen that for steep temperature gradients, nearly all transport of energy is by convection; for temperature gradients lower than the adiabatic value, the energy transport is entirely by radiation.

Thus for a case of convective equilibrium, we have the adiabatic condition holding in the star to a good degree of approximation:

$$\frac{1}{P}\frac{dP}{dr} = \frac{\gamma}{\rho}\frac{d\rho}{dr}.$$

In general, we can expect convection to be important near the centers of stars where thermonuclear energy generation, which often varies as a high power of the temperature, tries to establish a steep temperature gradient. Farther away from the center of a star, where energy generation is not important, the temperature gradient becomes less, and energy transport is by radiation. However, convection can also set in in a region where a major constituent of the gas is partly ionized. As a volume element of the gas rises in such a region, recombination takes place. This heats the volume element which continues to rise until the major constituent has become nearly all recombined. This process is responsible for establishing outer convection zones in stars where hydrogen is partly ionized.

4.2 Additional equations for the stellar interior

In order to determine the structure of a star, we need also to know some additional properties of the matter contained in the interior. One property is the equation of state of a perfect gas:

$$p_g = \frac{k}{H}\frac{\rho}{\mu}T,$$

where $k =$ Boltzmann's constant,

$H =$ mass of the hydrogen atom,

and μ = mean molecular weight of the matter, which is the inverse of the number of particles per unit atomic mass ($\mu = 1/2$ for hydrogen, $4/3$ for helium, and 2 for heavy elements).

The radiation pressure is given by

$$p_r = \frac{1}{3}aT^4.$$

The stellar opacity K is a very complicated function. In an accurate calculation, it is necessary to take into account, as detailed functions of photon frequency, all the scattering and absorbing processes which can take place in all the different kinds of atoms present. We shall not give a detailed theory of opacity but will content ourselves with mentioning the important physical processes which contribute to it.

Under the conditions usually present in the interiors of main sequence stars, the most important absorption process is photoelectric absorption in the heavier atoms present. Hydrogen and helium do not make appreciable contributions to the opacity in this process, since they are completely ionized.

A lesser source of opacity is the free-free transition, in which an electron moving in the continuum is excited to a higher state of the continuum.

At higher temperatures an important source of opacity becomes photon scattering by free electrons (Thomson scattering or the Compton effect).

Much work has been done with an approximation to the stellar opacity known as Kramers's formula:

$$K = \text{constant } \rho T^{-3.5}.$$

Here, the constant is proportional to the abundances of elements heavier than helium.

It may, therefore, be seen that the presence of heavier atoms in stellar interiors is very important from the point of view of the opacity. Higher abundances of heavy elements increase the opacity, which in turn raises the temperature gradients in the interior. The rate of energy generation at the center is therefore increased, and a star of given mass will be more luminous than one with a smaller content of heavy elements. We see here a reason why the main sequences of the globular clusters lie below that for the Population I stars.

In addition to a good opacity relation, it is also necessary to obtain from physical theory the dependence of the energy generation rate E on temperature, density, and chemical composition of the material, as well as any contributions due to gravitational contraction.

The four differential equations which apply at any point in a stellar interior, taken together with the additional relationships discussed here, allow a complete stellar model to be constructed, given only the total mass and distribution of chemical composition in the star. In practice such computations are very complex to carry out, and really good results in which stellar evolution has been followed through a series of models with changing chemical composition are only now starting to be obtained with electronic computers.

4.3 Electron degeneracy

The preceding discussion was based on the assumption that the perfect gas laws held to a good approximation in stellar interiors. This assumption fails at high densities owing to the onset of electron degeneracy.

Let us consider electrons which are confined to a unit volume. According to quantum theory, there are a limited number of states available to these electrons. The number of states available is

$$\left(\frac{2}{h^3}\right) 4\pi p^2 dp$$

in the momentum range p to $p + dp$. This allows two electrons to exist in each cell of phase space, one with each of the two possible spin orientations. The actual number of electrons in this momentum range will be less than or equal to the above quantity:

$$N(p)dp \leq \left(\frac{2}{h^3}\right) 4\pi p^2 dp.$$

Let us compare this with the number of electrons in the same momentum interval which we would have for a Maxwell distribution of velocities:

$$N(p)dp = \frac{N_e}{(2\pi mkT)^{3/2}} \, e^{-p^2/2mkT} \, 4\pi p^2 dp,$$

where N_e is the number of electrons per unit volume, and m is the mass of an electron. Now, at low momenta, the Maxwell distribution violates the Pauli principle if

$$\frac{N_e}{(2\pi mkT)^{3/2}} > \frac{2}{h^3}. \tag{4.3.1}$$

If $N_e = \frac{\rho}{\mu_e H}$, where μ_e is the mean molecular weight per electron, then (4.3.1) can be expressed in the form:

$$\rho > \frac{2(2\pi mkT)^{3/2}}{h^3}\mu_e H = 8.1 \times 10^{-9}\mu_e T^{3/2}. \tag{4.3.2}$$

For $\mu_e = 2$ and $T = 10^8$ °K, the Pauli principle would be violated for densities in excess of 1.62×10^4 g/cm^3. Serious violations would certainly occur at the much higher densities sometimes encountered in stellar interiors.

Therefore, at high densities, the Maxwell distribution is depleted of electrons with low momenta and enriched with electrons of higher momenta. At extremely high densities, a good approximation to complete degeneracy must set in, in which:

$$N(p) = \left(\frac{2}{h^3}\right) 4\pi p^2, \qquad (p \leq p_0)$$
$$\text{and} \quad N(p) = 0, \qquad (p > p_0) \tag{4.3.3}$$

in which p_0 is the Fermi threshold of the distribution (sometimes called the Fermi level, or the corresponding energy $p_0^2/2m$ is called the Fermi energy). It is given by

$$p_0 = \left[\frac{3h^3}{8\pi}N_e\right]^{1/3}.$$

Under conditions of interest in stellar interiors, electron degeneracy sets in at high densities under conditions such that the electrons move with high velocities which are a large fraction of the velocity of light. Hence relativistic mechanics must be used in the derivation of the equation of state of a degenerate electron gas. This has been done by Chandrasekhar [24; 25]. He obtains the following results for the pressure and density of a degenerate electron gas:

$$p = \frac{\pi m^4 c^5}{3h^3}f(x), \tag{4.3.4}$$

31

$$\text{where} \quad f(x) = x(2x^3 - 3)(x^2 + 1)^{1/2} + 3\operatorname{arcsinh} x,$$

$$\text{and} \quad x = \frac{p_0}{mc}; \tag{4.3.5}$$

$$\rho = \frac{8\pi m^3 c^3}{3h^3} \mu_e H x^3.$$

(4.3.4) and (4.3.5) represent parametrically the equation of state of a highly degenerate electron gas. The pressure in this gas is much larger than in a gas obeying the perfect gas laws. Such additional pressure will be referred to for convenience as electron degeneracy pressure.

Highly degenerate matter in equilibrium under its own gravitation has some very interesting properties. The temperature of the matter is irrelevant as long as it is not so high that it destroys the good approximation to complete degeneracy. However, the degenerate matter is a good thermal conductor, so that the temperature distribution in the matter must be very flat. It turns out that there is an upper limit to the mass of such a completely degenerate body; the degeneracy pressure can support the weight of the ions and electrons provided that weight is not too large. The radius of the degenerate body approaches zero as this limiting mass is approached. The limiting mass is:

$$M < \frac{5.756}{\mu_e^2} \text{ solar masses,}$$

or, for $\mu_e = 2$, $M < 1.44$ solar masses. There are some corrections which should be made which somewhat reduce the value of the limiting mass given above. White dwarf stars appear to be highly degenerate bodies of the sort described to a good degree of approximation. The white dwarfs contain no energy sources and gradually cool off at a constant radius as their internal heat content is radiated to space from the thin, nondegenerate atmosphere.

One further property of degenerate matter should be mentioned owing to its importance in the subsequent discussion. Let us consider a volume element of the degenerate matter and let this volume element be suddenly compressed. The Fermi threshold of the electrons is raised at the higher density which results from the compression. The internal energy stored in the electrons is increased, and this energy can only come at the expense of the kinetic energy of the ions present which are not degenerate. Accordingly, the ionic temperature is reduced during a compression. Similarly, when a degenerate gas expands the ionic temperature increases. This behavior is opposite to that of an ordinary gas. If the temperature of the ions is high enough for the ions to take part in thermonuclear reactions, then the degenerate gas is very unstable.

Chapter **5**

Thermonuclear Reaction Rates

Although we have seen that electrons can form a degenerate gas in stellar interiors, we do not expect that the ions will become degenerate, owing to their much larger masses. Therefore, the ions will have a Maxwellian distribution of velocities under all conditions of interest. The tail of the Maxwell distribution decreases in an exponential manner, but nevertheless, when considering thermonuclear reactions we are particularly interested in velocities in the far tail of the distribution, corresponding to particle energies often five to ten times kT. The reason for this is that nuclear reactions take place only when colliding charged particles can penetrate their mutual Coulomb barrier far enough for the particles to come within the range of the nuclear forces of each other. The probability for this penetration varies exponentially as the inverse of the velocity. The products of these two exponential factors has a sharp peak in the far tail of the Maxwell distribution.

5.1 Nonresonant thermonuclear reactions

Most nuclear reactions have resonances; at certain values of the bombarding energy, the reaction cross section is much higher than at most other bombarding energies. We shall first be concerned with the case in which the thermonuclear bombardment energy region is far from the nearest resonance. However, the thermonuclear reaction rates will be formulated generally [26].

The reaction rate r is equal to the product of the number of possible pairs of reacting

34

particles and to the probability that a reaction will take place between any reacting pair. For reactions between unlike particles it is

$$r = n_1 n_2 \langle \sigma v \rangle,$$

and for reactions between like particles it is

$$r = \frac{1}{2} n^2 \langle \sigma v \rangle.$$

Here $n_1, n_2,$ and n are the numbers of reaction particles per cm^3. The quantity $\langle \sigma v \rangle$ is an appropriate average of the product of the reaction cross section and the relative velocity of the reacting particles. It depends only on the temperature of the gas and the nuclear properties of the reaction.

Let the particles have velocity distribution functions $n_1(v)$ and $n_2(v)$. Then

$$r = \int \int n_1(\vec{v_1}) \, n_2(\vec{v_2}) \, |\vec{v_1} - \vec{v_2}| \, \sigma(\vec{v_1} - \vec{v_2}) \, d^3\vec{v_1} \, d^3\vec{v_2},$$

where the volume integrals cover both velocity spaces. Putting in Maxwellian velocity distributions, we obtain

$$r = n_1 n_2 \left(\frac{m_1}{2\pi kT}\right)^{3/2} \left(\frac{m_2}{2\pi kT}\right)^{3/2} \int \int d^3\vec{v_1} \, d^3\vec{v_2} \, |\vec{v_1} - \vec{v_2}| \, \sigma(\vec{v_1} - \vec{v_2})$$

$$\times \exp\left(-\frac{\frac{1}{2}m_1 v_1^2 + \frac{1}{2}m_2 v_2^2}{kT}\right),$$

where m_1 and m_2 are the masses of the two particles.

We reduce the integral by changing to the relative velocity $\vec{v} = \vec{v_1} - \vec{v_2}$ and to the center of mass velocity

$$\vec{V} = \frac{m_1 \vec{v_1} + m_2 \vec{v_2}}{m_1 + m_2}.$$

The integral over $d^3\vec{V}$ and over the solid angle in $d^3\vec{v}$ can be carried out immediately. This gives

$$r = n_1 n_2 4\pi \left(\frac{m_1 m_2}{2\pi kT(m_1 + m_2)}\right)^{3/2} \int_0^\infty dv \, v^3 \, \sigma(v) \exp\left[-\left(\frac{m_1 m_2}{m_1 + m_2}\right) \frac{v^2}{kT}\right]. \quad (5.1.1)$$

At low bombarding energies and far away from the resonances, nuclear reaction cross sections can be expressed in the form:

$$\sigma = \frac{S}{E} \exp\left(-\frac{2\pi e^2 Z_1 Z_2}{\hbar v} \right),$$

where E is the relative energy which corresponds to the velocity v, e is the electric charge, and Z_1 and Z_2 are the atomic numbers of the colliding particles. We write $E = \frac{1}{2}mv^2$, where $m = \frac{m_1 m_2}{m_1 + m_2}$. S is a slowly varying function of the bombarding energy which is considered to be constant in the present formulation of problem. The reaction rate becomes:

$$r = 2n_1 n_2 \left(\frac{2}{\pi m} \right)^{1/2} \frac{S}{(kT)^{3/2}} \int_0^\infty dE \ \exp - \left(\frac{E}{kT} + \frac{B}{E^{1/2}} \right),$$

where $\frac{B}{E^{1/2}} = \frac{2\pi e^2 Z_1 Z_2}{\hbar v}$. The integrand is sharply peaked and may be approximated by a Gaussian function. When this is done we get the following good approximation:

$$\int_0^\infty dx \ \exp -(x + ax^{-1/2}) \approx 2 \left(\frac{1}{3}\pi \right)^{1/2} \left(\frac{1}{2}a \right)^{1/3} \exp \left[-3 \left(\frac{1}{2}a \right)^{2/3} \right].$$

This completes the derivation of the thermonuclear reaction rate for some rearrangement to make the formulas convenient to use. We express the number density of particles in terms of the gas density by the reaction

$$n_1 = \frac{N_0 \rho x_1}{A_1},$$

where N_0 is Avogadro's number, A_1 is the mass number of the particle, and x_1 is the concentration by weight of the particles of type 1 in the gas. We shall also express our energies in the center of mass system of the bombarding particles. Then p, the number of reactions per second per nucleus of type 2 is

$$p = \frac{434 \rho x_1 (A_1 + A_2) S}{A_1^2 A_2 Z_1 Z_2} \tau^2 e^{-\tau} \text{s}^{-1}, \tag{5.1.2}$$

$$\text{where} \quad \tau = 42.48 \left[\frac{Z_1^2 Z_2^2 A_1 A_2}{A_1 + A_2} \right]^{1/3} T^{-1/3},$$

$$\text{and} \quad S = \sigma E \exp\left[988 Z_1 Z_2 \left(\frac{A_1 A_2}{(A_1 + A_2) E} \right)^{1/2} \right]. \tag{5.1.3}$$

Here σ is the cross section for the reaction in barns, E is the bombarding energy in the center of mass in electron volts, and the temperature T is in units of 10^6 °K.

The quantity S can sometimes be determined for a specific reaction by measuring the reaction cross section at the lowest possible bombarding energy and putting the measured value in equation (5.1.3). In other cases S must be calculated from resonance theory or from the statistical properties of nuclear resonances. This is discussed in greater detail in the subsequent sections.

Most reactions take place in the vicinity of the energy about which the approximated integrand is sharply peaked. This peak occurs at an energy

$$E_m = 1220 \left[\frac{Z_1^2 Z_2^2 A_1 A_2 T^2}{A_1 + A_2} \right]^{1/3} \quad \text{electron volts.}$$

The full width at half maximum of the integrand is

$$\Delta E = 15.2 (T E_m)^{1/2} \quad \text{electron volts.}$$

This peak in the integrand will be called the "Gamow peak."

5.2 Resonant thermonuclear reactions

We now consider the case where there is a nuclear resonance close to the Gamow peak. In most cases of this sort in which we shall be interested, the width of the resonance is very small compared to the width of the Gamow peak (here and in general the term "width" will be taken to be the full width at half-maximum of a peak). In this case, equation (5.1.1) can be written:

$$r = 4\pi n_1 n_2 \left(\frac{m_1 m_2}{2\pi k T(m_1 + m_2)} \right)^{3/2} v_r^3 \exp\left[-\left(\frac{m_1 m_2}{m_1 + m_2} \right) \frac{v_r^2}{kT} \right] \int_0^\infty \sigma(v) dv,$$

where v_r is the relative velocity of the particles corresponding to the resonance. We see that the integral now is just the integrated cross section under the nuclear resonance.

In a very refined treatment, nuclear cross sections would be expressed in terms of the Breit-Wigner many-level formula. However, for most astrophysical purposes, it is good enough to use the Breit-Wigner single-level formula, particularly in the vicinity of a resonance peak such as we are interested in here. The Breit-Wigner formula for the cross section of a reaction in which a particle a enters the nucleus and a particle b emerges from it is:

$$\sigma(a,b) = \frac{\pi \lambda_a^2 g \Gamma_a \Gamma_b}{(E_r - E)^2 + \Gamma^2/4}, \tag{5.2.1}$$

where λ_a is the de Broglie wave length of particle a divided by 2π; Γ_a, Γ_b, and Γ are the partial widths of particles a and b and the total width of the level, respectively; E is the bombarding energy; and E_r is the energy of the resonance. The statistical factor g is given by

$$g = \frac{2J + 1}{(2s + 1)(2I + 1)},$$

where s is the spin of the bombarding particle, I is the spin of the target nucleus, and J is the spin of the resonant level in the compound nucleus.

The integrated cross section is given to a good approximation by

$$\int_0^\infty \sigma(a,b) dE \approx 2\pi^2 \lambda_a^2 g \frac{\Gamma_a \Gamma_b}{\Gamma},$$

where λ_a^2 must be evaluated at E_r.

We write a subscript l for particle a and a subscript γ for particle b (to indicate the fact that the resonance width is not likely to be smaller than the width of the Gamow peak unless the emergent particle is a photon). Then a resonant thermonuclear reaction rate, where again the temperature T is in units of 10^6 °K, becomes:

$$p = 4.81 \times 10^9 \frac{\rho x_1 g}{A_1 T^{3/2}} \frac{\Gamma_l \Gamma_\gamma}{\Gamma} \left(\frac{A_1 + A_2}{A_1 A_2} \right)^{3/2} \exp \left(\frac{-0.0116\, E_r}{T} \right) \mathrm{s}^{-1}. \tag{5.2.2}$$

E_r is the resonance bombarding energy in electron volts in the center of mass system.

Resonant thermonuclear reaction rates are usually higher than nonresonant rates by a factor of something like 10^5. This indicates the extreme importance of knowing whether there are resonances in the thermonuclear energy region when considering the astrophysical important of specific nuclear reactions.

In most reactions of interest the total width $\Gamma = \Gamma_l + \Gamma_\gamma$. Usually one of these two partial widths is much smaller than the others. In such a case the ratio $\frac{\Gamma_l \Gamma_\gamma}{\Gamma}$ can be replaced by the smaller partial width, and the larger partial width need not be known at all.

5.3 Application of the Wigner-Eisenbud theory

Unless nuclear cross sections can be measured in the laboratory at energies close to the Gamow peak, and unless the closest resonance level is far from the Gamow peak, then the cross section at the Gamow peak must be determined using nuclear theory. A refined formulation of nuclear reaction theory has been given by Wigner & Eisenbud [27], often called the dispersion theory of nuclear reactions. According to dispersion theory the partial width of a particle can be written in the form

$$\Gamma = 2kP\gamma^2, \tag{5.3.1}$$

where k is the wave number of the incident particle,

$$k = 2.19 \times 10^9 \left(\frac{A_1 A_2 E}{A_1 + A_2} \right)^{1/2} \mathrm{cm}^{-1}, \tag{5.3.2}$$

P is the barrier penetrability of the particle, and γ^2 is the reduced width of the particle in the level under consideration. In this theory the configuration space is divided into an internal region, to which all specifically nuclear effects are restricted, and an external region in which the system is composed of two particles interacting only through the Coulomb field. The barrier penetrability is

$$P = \frac{1}{F^2 + G^2}, \tag{5.3.3}$$

where F and G are the regular and irregular solutions of the Schrödinger equation for a particle in a Coulomb field. For the low energies which are involved here,

$$G^2 \gg F^2.$$

The penetrability can be evaluated using the low energy approximation given by Yost, Wheeler & Breit [28]. It is necessary to use two functions of the bombarding energy, measured as before in electron volts:

$$\rho = kR, \tag{5.3.4}$$

$$\text{and } \eta = 157.4 \ Z_1 Z_2 \left(\frac{A_1 A_2}{(A_1 + A_2)E} \right)^{1/2}. \tag{5.3.5}$$

Here, R is the radius of the boundary between the internal and external regions in the configuration space. In principle, it should not matter what value is used for the radius of this boundary surface, provided it is taken large enough to avoid nuclear effects in the external region. However, in practice it is desirable to use as small a value as possible for the nuclear interaction distance, and it is customary to take it equal to the sum of the radii of the two interacting particles:

$$R = 1.45 \times 10^{-13} \ (A_1^{1/3} + A_2^{1/3}) \ \text{cm}.$$

As we have seen, the barrier penetrability for particles with angular momentum L is G_L^{-2}. This can be written in the form

$$\frac{1}{G_L^2} = \frac{\rho^{2L}}{D_L^2 \theta_L^2},$$

$$\text{where } \frac{1}{D_L^2} = \frac{2^{2L}}{[(2L)!]^2} \left[(L^2 + \eta^2)([L-1]^2 + \eta^2) \dots (1^2 + \eta^2) \right] \left[\frac{2\pi\eta}{e^{2\pi\eta} - 1} \right].$$

$$\text{We write } q = \frac{2^{2L}}{[(2L)!]^2} \left[(L^2 + \eta^2)([L-1]^2 + \eta^2) \dots (1^2 + \eta^2) \right].$$

Under the conditions of interest $\exp(2\pi\eta)$ is large compared to unity, and hence we can write to a good approximation

$$\frac{1}{G_L^2} \approx \frac{2\pi \eta q \rho^{2L}}{\theta_L^2} \exp(-2\pi \eta),$$

for $L = 0$, $q = 1$. We may evaluate the function θ_L in the low energy limit which is

$$\theta_L = -\left[\frac{2}{(2L)!}\right]\left(\frac{x}{2}\right)^{2L+1} K_{2L+1}(x),$$

where $x = (8\rho\eta)^{1/2}$ and $K_{2L+1}(x)$ is the modified Bessel function of the second kind.

Teichmann & Wigner [29] have found a practical sum rule limit for the reduced width γ^2 whose order of magnitude is (dropping a factor of 1.5 from Teichmann & Wigner's formula):

$$\gamma^2 \lesssim \frac{\hbar}{mR}, \tag{5.3.6}$$

where, as before, m is the reduced mass of the colliding particles, and R is the radius of the boundary between the internal and external regions of configuration space. Lane [30]* has calculated reduced widths for nucleons in single particle states in light nuclei and finds values close to the upper limit given by the above inequality. Therefore, $\hbar/mR$ will be called the single particle limit of the reduced widths.

According to a modern view of nuclear level structure, a single nucleon with a given angular momentum can interact with a nucleus to form a single particle state about once in every 15 Mev of excitation energy. In practice, the wave functions for these single particle states mix into the wave functions of all the nuclear states in the vicinity of the single particle level excitation energy. The reduced width for the single particle in any given nuclear state thus depends on the amount to which the single particle wave function is mixed into the state. The smaller the spacing between the nuclear levels, the larger the number of levels competing for the single particle wave function, and the smaller the reduced width is expected to be for the single particle in any given level. If D is the energy spacing between particles of the same spin and parity, then γ^2/D is called the nuclear strength function for the interaction of the given individual particle with the nucleus. We expect the strength function to be large near the natural positions of the

---

*Editor's Note: Results in this report were later accepted in a peer-reviewed journal [31].

single particle levels in the nucleus and to be small far away from the natural position. If we do not know where the natural position of the single particle levels are, then we can assume that the strength function has been uniformly spread out over all excitation energies. In this case it has the value [32]:

$$\frac{\gamma^2}{D} \approx 2 \times 10^{-14} \text{ cm.} \tag{5.3.7}$$

Of course, one cannot expect either of these methods of estimating the strength function necessarily to give a good value of the reduced width for an individual level, since there are large differences in the degree to which the single particle wave function will mix into the wave functions for different states. Porter & Thomas [33] have argued on reasonable grounds that the distribution of reduced widths will be given approximately by the function:

$$P(x) = \frac{\exp(-\frac{1}{2}x)}{(2\pi x)^{1/2}}, \tag{5.3.8}$$

where $x = \gamma^2/\langle\gamma^2\rangle$.

This distribution function indicates that most reduced widths are considerably smaller than the average, and a few are much larger than the average. Such a great variability of individual reduced widths should always be borne in mind when estimating unknown physical parameters for astrophysical purpose. One should certainly not draw any strong conclusions which depend upon one or more reduced widths being close to the average value. It should also be emphasized that the considerations in this section apply to incident or outgoing particles which can be absorbed or emitted in only a single channel, which means that there is only one state into which the nucleus and particle can separate. When a particle is emitted into many channels, then the sum of the reduced widths for each channel starts to approach the sum of the average reduced widths.

We now give for convenience the contribution that a distant level will make to the nonresonant parameter S, assuming that there is no interference between levels and that the contributions of all the distant levels add linearly. We assume as before that the cross section $\sigma(a,b)$ for the reaction is given by the single-level Breit-Wigner formula.

Then

$$S = 2.84 \times 10^{18} \, \frac{q\rho^{2L}}{\theta_L^2} \, \frac{Z_1 Z_2 (2J+1)}{(2s+1)(2I+1)} \left[\frac{\gamma_a^2 \Gamma_b}{(E-E_r)^2 + 1/4\Gamma^2} \right] \text{ev barns.}$$

If there is serous interference, then the quantity in the square brackets would have to be replaced by a similar quantity obtained from the Breit-Wigner many-level formula. In cases where the reduced widths γ_a^2 and γ_b^2 are not known, then we can get an estimate of the contribution to S by estimating them on statistical grounds. The above expression for S should be evaluated at the Gamow peak energy E_m.

Two further correction factors to the thermonuclear reaction rates will be mentioned briefly. They are small and of importance only for very accurate work.

Salpeter [34] has considered the errors which were made in the Gaussian approximation to the integrand of (5.1.1). He finds that the thermonuclear reaction rates should be multiplied by a correction factor:

$$F_\tau = 1 + \frac{5}{12\tau} - \frac{35}{288\tau^2} + \cdots$$

Corrections due to electron screening have been considered by Schatzman [35; 36; 37], Keller [38], and Salpeter [39]. Thermonuclear reaction rates are slightly increased because the penetration of the Coulomb barriers is slightly assisted by the presence of electrons which neutralize the Coulomb field at large distances from the nucleus. The formulation of the electron-shielding corrections is complex, and the resulting correction factors are usually not much greater than unity. No further consideration will be given here to these corrections.

In many cases, it is useful to be able to speak of the temperature sensitivity of a thermonuclear reaction. Let us write

$$p = p_0 T^n.$$

Then the power n is obtained from the relation

$$n = \frac{d \log p}{d \log T}.$$

For nonresonant reactions it follows that:

$$n = \frac{1}{3}(\tau - 2).$$

For resonant reactions the power takes the form:

$$n = \frac{0.0116 E_r}{T} - \frac{3}{2}.$$

Chapter 6

Hydrogen Thermonuclear Reactions

In this Chapter, we consider those hydrogen thermonuclear reactions which are of interest in the study of stellar interiors and of the changes in chemical abundances which can be produced by them. Those values of the nuclear reaction constants which appear to be the best available in the literature have been listed in Table 6.1. No attempt has been made to improve these constants except in the case of the proton-proton reaction, which is of particular importance in the construction of stellar models, and also in the case of the $N^{14}(p,\gamma)O^{15}$ reaction, which is of similar importance. It turns out that none of the hydrogen thermonuclear reactions listed have any known nuclear resonances in the vicinity of the Gamow peak; hence the nonresonant rate formula (5.1.2) applies to these reactions. This equation can be written in the form

$$p = A(\rho x_1)T^{-2/3} \exp(-B/T^{1/3}) \text{ s}^{-1},$$

$$\text{where} \quad A = \frac{434 \ (A_1 + A_2)S}{A_1^2 A_2 Z_1 Z_2}(\tau T^{1/3})^2,$$

$$\text{and} \quad B = \tau T^{1/3}.$$

We shall discuss these reactions not in the order of increasing Z and A as

Table 6.1

Reaction	S (e.v. barns)	A	B	Note	Reference
$H^1(p, \beta^+\nu)H^2$	3.12×10^{-19}	3.3×10^{-13}	34.7	1	[34]
$H^2(p, \gamma)He^3$	7.8×10^{-2}	7×10^4	37.2		[40]
$H^2(d, n)He^3$ $\}$ $H^2(d, p)H^3$	9.6×10^4	3.6×10^{10}	42.6	2	[40]
$Li^6(p, \alpha)He^3$	5×10^6	6×10^{12}	84.1		[8]
$Li^7(p, \alpha)He^4$	1.0×10^5	1.2×10^{11}	84.7		[8]
$Be^9(p, d)Be^8$ $\}$ $Be^9(p, \alpha)Li^6$	3×10^7	3×10^{13}	103.6		[8]
$B^{10}(p, \alpha)Be^7$	2×10^7	2.5×10^{13}	120.6		[8]
$B^{11}(p, \alpha)Be^8$	1×10^8	1.2×10^{14}	121.0		[8]
$C^{12}(p, \gamma)N^{13}$	1.2×10^3	1.7×10^9	136.5		[41]
$C^{13}(p, \gamma)N^{14}$	6.1×10^3	9×10^9	136.9		[41]
$N^{14}(p, \gamma)O^{15}$	$\sim 3 \times 10^3$	5×10^9	152.8	3	[41]
$N^{15}(p, \alpha)C^{12}$	1.1×10^8	1.7×10^{14}	153.1		[41]
$N^{15}(p, \gamma)O^{16}$	1.1×10^4	1.7×10^{10}	153.1		†
$O^{16}(p, \gamma)F^{17}$	$> 10^2$	$> 1.5 \times 10^8$	166.7	4	[8]
$O^{17}(p, \alpha)N^{14}$	2×10^5	3×10^{11}	167		[8]
$O^{18}(p, \alpha)N^{15}$	1×10^8	1.5×10^{14}	167		[8]
$F^{19}(p, \alpha)O^{16}$	1×10^9	1.5×10^{15}	181		[8]
$Ne^{20}(p, \gamma)Na^{21}$	1.2×10^4	1.8×10^{10}	194		[42]
$Ne^{21}(p, \gamma)Na^{22}$	$> 5.6 \times 10^3$	$> 1 \times 10^{10}$	194	5	[42]
$Ne^{22}(p, \gamma)Na^{23}$	$\geq 5.5 \times 10^5$	$\geq 1 \times 10^{12}$	194		[42]
$Na^{23}(p, \gamma)Mg^{24}$	3×10^6	3×10^{12}	208		[42]
$Na^{23}(p, \alpha)Ne^{20}$	$\gg 10^7$	$\gg 10^{13}$	208		[42]
$He^3(\alpha, \gamma)Be^7$	0.6	1.7×10^5	128.0	6	[34]
$He^3(He^3, 2p)He^4$	1×10^6	4×10^{11}	122.5	7	[34]

† A. G. W. Cameron, unpublished.

Notes to Table 6.1

1. The basic proton-proton reaction rate has been re-evaluated as described in the text (Salpeter's result has been multiplied by a factor of 0.72). The error in the reaction rate is now estimated to be nine percent, compounded from a five percent error in the orbital matrix element, a three percent error in the beta-decay f-value, and a seven percent error in the square of the Gamow-Teller coupling constant. The values of S and A in the table are evaluated at $T = 15 \times 10^6$ °K; at other temperatures, the values should be corrected as indicated in the text. It should also be noted that the values of S and A refer to a single reaction; to obtain the reaction rate per cubic centimeter, it is necessary to multiply by one-half times the number of protons per cubic centimeter. In this respect, the values of S and A differ from those given in reference [8], where the factor of one-half has been included in S and A.

2. The concentration of deuterium by weight is used in the formula for p in this case. An additional factor of one-half should be included in the reaction rate per cubic centimeter.

3. The reaction rate is less than that of reference [41] by a factor of 10 for reasons discussed in the text.

4. The values of S and A from reference [8] are taken to be lower limits as discussed in the text.

5. S and A are evaluated at $T = 50 \times 10^6$ °K. There are fairly large corrections at other temperatures.

6. Here p contains the concentration by weight of He^4.

7. Here p contains the concentration by weight of He^3. An additional factor of one-half should be included in the reaction rate per cubic centimeter.

they are listed in Table 6.1, but in the order of importance in a gas in which the temperature and density are progressively increased.

6.1 Deuterium reactions

When a star contracts from interstellar space and the central temperature increases, the first thermonuclear reactions to set in are those which destroy the deuterium. The ratio of abundances (by weight) of deuterium and hydrogen in the earth and meteorites is 1.5×10^{-4}; hence the deuterium does not provide one of the major sources of energy for the stars. Nevertheless, the energy obtained from the deuterium reactions is comparable to the energy obtained from the gravitational contraction of the star, and the deuterium consumption stage must have a profound influence on the details of the stellar model in the contraction stage. This is also the stage in which the planets are believed to have formed in the solar system. Hence it is probably necessary to obtain a good model for the deuterium-burning sun in order to construct a good theory of planetary formation.

Deuterium can be destroyed both by direct proton capture and in reactions with other deuterium nuclei. The reactions which we must consider are

$$H^2(p, \gamma)He^3, \tag{6.1.1}$$

$$H^2(d, n)He^3, \tag{6.1.2}$$

$$H^2(d, p)H^3. \tag{6.1.3}$$

It may be seen in Table 6.1 that the nuclear reaction constant A is much smaller for the $H^2(p, \gamma)He^3$ reaction than for the d+d reactions combined. However, this is more than offset by the difference in the abundances of hydrogen and deuterium. Salpeter [8] has discussed the behavior of these reactions.

In a red dwarf star on the main sequence, with a luminosity perhaps 10^{-4} of that of the sun, the deuterium will be destroyed at a temperature of about 0.5×10^6 °K. In the sun it was destroyed at a temperature of about 1×10^6 °K, and in a very hot main sequence star of Population I the destruction will occur at a temperature of about 2×10^6 °K. At these three temperatures, the d-d and d-p reaction rates are equal for concentration ratios (X_{H^2}/X_{H^1}) of 2×10^{-3}, 5×10^{-4}, and 2×10^{-4} respectively. We see that the $H^2(p, \gamma)He^3$ reaction predominates, except in the hottest stars, and even in the hottest stars, the d-d reactions provide effective competition only at the beginning of the deuterium destruction when the deuterium concentration has nearly its initial value. Nevertheless some neutrons and some tritium will be produced by d-d reactions.

Reactions (6.1.2) and (6.1.3) have nearly the same yield. The tritium production decays into He^3 with a half-life of 12 years, which is much faster than the tritium can be destroyed by thermonuclear reactions. The neutrons produced are ejected with energies of 2.5 Mev. They then exist in a medium nearly all hydrogen for which the scattering cross sections are larger than the neutron absorption cross sections. Therefore, the neutrons are slowed down until they are in thermal equilibrium with their surroundings, which gives neutron kinetic energies in the vicinity of 100 electron volts. At this energy, hydrogen is so much more abundant than any other element, and the neutrons will be captured by it unless some other nucleus has a sufficiently large absorption cross section to offset the abundance disadvantage. The only possibility which seems to exist is the reaction $He^3(n, p)H^3$, which has a cross section of 81 barns at 100 electron volts. This reaction will absorb most of the neutrons only if the ratio of the concentrations by mass of He^3 and hydrogen is larger than 2×10^{-4}, which appears somewhat unlikely.

The He^3 which is formed in these reactions is stable under the conditions being considered.

6.2   Lithium, beryllium, and boron reactions

After the exhaustion of deuterium, a star will continue contracting, and its central temperature will continue rising. In succession, the elements lithium, beryllium, and boron will be destroyed. The abundances of these elements are so small that the energy produced during their destruction is very small compared to the energy released in the gravitational contraction.

The lithium isotopes are destroyed when the temperature reaches about 3×10^6 °K. The reactions are:

$$Li^6(p, \alpha)He^3, \qquad (6.2.1)$$

and $\qquad$ $Li^7(p, \alpha)He^4$. $\hfill$ (6.2.2)

The He^3 produced in (6.2.1) is still stable at $T = 3 \times 10^6$ °K.

There are two exothermic reactions by which Be^9 can be destroyed. These are

$$Be^9(p, d)Be^8(2\alpha), \hfill (6.2.3)$$

followed by $\qquad$ $H^2(p, \gamma)He^3$, $\hfill$ (6.1.1)

and $\qquad$ $Be^9(p, \alpha)Li^6$, $\hfill$ (6.2.4)

followed by $\qquad$ $Li^6(p, \alpha)He^3$. $\hfill$ (6.2.1)

These reactions destroy beryllium at a temperature of about 4×10^6 °K. The He^3 formed is still stable at this temperature.

The two boron isotopes are also destroyed by reactions in which particles are emitted. These are:

$$B^{10}(p, \alpha)Be^7, \hfill (6.2.5)$$

followed by $\qquad$ $Be^7(e, \nu)Li^7$ (electron capture), $\hfill$ (6.2.6)

and by $\qquad$ $Li^7(p, \alpha)He^4$, $\hfill$ (6.2.2)

and $\qquad$ $B^{11}(p, \alpha)Be^8(2\alpha)$. $\hfill$ (6.2.7)

These reactions take place when the temperature reaches 6 or 7×10^6 °K.

It should be noticed that all of the reactions given in this section produce He^3 and He^4 nuclei as their final products. Both of these nuclei are stable in the temperature range under consideration, although, as we shall see, the He^3 is destroyed when the temperature rises a little more.

6.3 The proton-proton chain

When less massive stars reach the main sequence, they stay at that particular point in the Hertzsprung-Russell diagram for a very long time while they convert hydrogen into helium in the central regions. The central temperature at which this occurs varies from about 8×10^6 °K in a faint red dwarf star to about 15×10^6 °K in the sun. In this temperature range, the conversion of hydrogen takes place almost entirely by the reactions in the proton-proton chain. These are:

$$H^1(p, \beta^+\nu)H^2, \tag{6.3.1}$$

followed by $\qquad H^2(p, \gamma)He^3, \tag{6.1.1}$

and by $\qquad He^3(He^3, 2p)He^4 \tag{6.3.2}$

or to a smaller extent followed by

$$He^3(\alpha, \gamma)Be^7, \tag{6.3.3}$$

$$Be^7(e, \nu)Li^7, \tag{6.2.6}$$

and $\qquad Li^7(p, \alpha)He^4. \tag{6.2.2}$

It may be seen in Table 6.1 that reaction (6.3.3) is considerably less likely than reaction (6.3.2). Therefore, two proton-proton reactions are needed for the formation of each alpha-particle.

Reaction (6.3.1) is extremely slow because a beta transformation must take place during the collision of two protons, which is a very improbable process. Thus (6.3.1) does not take place at an appreciable rate until the temperature and density of the gas become high enough that the protons spend an appreciable fraction of time in collisions in which the protons penetrate their mutual potential barrier far enough to come within the range of the nuclear forces of each other. There is no hope of ever observing this reaction in the laboratory. Nevertheless, the two-body problem in nuclear physics is quite accurately solved at low energies, and the theory of beta decay, while needing reformulation after the discovery of nonconservation of parity in weak interactions, is sufficiently well developed for such simple systems that the thermonuclear reaction rate for (6.3.1) can be calculated quite accurately.

This calculation was first performed by Bethe & Critchfield [43]. They have written the cross section σ for the combination of two protons of relative velocity v as:

$$\sigma = gf(W)v^{-1}\left|\int \psi_p\psi_d d\tau\right|^2 |M_{sp}|^2,$$

where g is the appropriate beta-decay coupling constant, $f(W)$ is the beta-decay f-function evaluated for the endpoint W of the positrons emitted in the decay, ψ_p is the wave function of the protons normalized per unit density at infinity, ψ_d is the wave function of the ground state of the deuteron, and M_{sp} is the spin part of the matrix element (omitted in the original calculation). The two protons undergo an s-wave interaction at low energies in which their spins are oppositely oriented; this forms a state of zero total angular momentum. The ground state of the deuteron is an s state in which the neutron and proton spins are lined up in the same direction; this state has a total angular momentum of unity. The beta transition between these states involves a unit change of spin which is allowed by Gamow-Teller selection rules. Hence g is the Gamow-Teller coupling constant (not expressed in the usual units). The factor $|M_{sp}|^2$ takes care of the summation over spin states in the transition and of the fact that either proton can turn into a neutron; it is equal to three-halves.

The matrix element has been evaluated very accurately by Frieman & Motz [44]. They used accurate explicit wave functions ψ_p and ψ_d based on specific assumptions about the potential shape of nuclear forces with the constants chosen to fit low energy experimental data for the two-nucleon system. However, in such a calculation it is very difficult to evaluate the inaccuracy in the final result due to the uncertainty in the potential shape and experimental errors. Therefore, Salpeter [34] has made an approximate re-evaluation of the matrix element using the theory of effective range, from which the effect of present uncertainties in the theory of nuclear forces can be derived very simply. Salpeter's calculation is less accurate because of the neglect of tensor forces and of certain approximations, but Salpeter's errors can be used with the calculations of Frieman & Motz [44] for the matrix element. Salpeter estimates that the matrix element has a probable error of five percent.

Salpeter evaluated the f-function using accurate approximations given by Feenberg & Trigg [45]. He estimates that this is subject to an error of three percent. Salpeter made an estimate of the Gamow-Teller coupling constant which was subject to an error of twenty percent, this being by far the greatest error in the calculation. It is now possible to use a more accurate value for this coupling constant. J. M. Robson (private communication) has recently obtained estimates of the Fermi scalar coupling constant using the more accurately measured half-lives and reaction energies for 0 to 0 transitions. He has also calculated the ratio of the Gamow-Teller tensor coupling constant to the Fermi constant from the ft value of the neutron, from the ft values of mirror nuclei, and from the angular correlation in the beta decay of the neutron. From this procedure he obtains (in Salpeter's units)

$$g = (5.38 \pm 0.36) \times 10^{-4} \text{ s}^{-1}.$$

Hence Salperer's reaction rate must be multiplied by a factor of 0.72. The error in the final result due to all sources is nine percent. The reaction rate is:

$$p = 2.48(1 \pm 0.09) \times 10^{-16} \, px_1 \tau^2 e^{-\tau} \left[1 + \frac{5}{12\tau} + \ldots \right] \left[1 + 0.054 \left(\frac{T}{15} \right)^{2/3} \right] \text{s}^{-1}.$$

In this expression a factor of one-half included by Salpeter has been removed. To obtain the reaction rate per cubic centimeter, it is necessary to multiply p by one-half times the number of protons per cubic centimeter.

6.3.1 Supplementary Notes: The proton-proton reaction

When the re-evaluation of the beta-decay matrix element in reaction (6.3.1) was made, it was still believed that the Fermi transitions involved scalar interactions and the Gamow-Teller transitions involved tensor interactions. Now it is believed that these interactions are vector and axial vector, respectively. These changes do not affect the revised value of the matrix element. It is interesting to note that Burbidge, Burbidge, Fowler, & Hoyle [1] have also derived a correction factor for the matrix element in close agreement with that found here (a correction factor of 0.71 compared to 0.72).

6.3.2 Supplementary Notes: The He$^3(\alpha, \gamma)$Be^7 reaction

Holmgren & Johnston [46] have measured the cross section for this reaction from 0.47 to 1.32 Mev. They find their points can be fitted by the expression $\sigma = 0.5E^{3.25}$ (microbarns). This expression is well fitted by the standard formula (equation (5.1.3)) with $S = 700$ ev barns (H. D. Holmgren, private communication*). It should be noted that this cross section is about 1000 times larger than the value given in Table 6.1 for this reaction (6.3.3). This probably implies that the reaction proceeds by a direct, non-resonant capture process similar to that described later on p. 67. In this case it would be necessary that there be a large reduced α-particle (or He^3) width in both the ground and first excited states of Be^7, since each participates about equally in the capture process.

Dr. Fowler (private communication) has remarked that the value of the cross section constant S for the reaction (6.3.2), He^3(He^3, 2p)He^4, should be about double that given in Table 6.1 owing to corrections for electron shielding.

The measurement of Holmgren & Johnston has tremendously important astrophysical implications. At higher temperatures, the He$^3(\alpha, \gamma)$Be^7 reaction will complete the proton-proton chain rather than the He^3(He^3, 2p)He^4 reaction. In the former case, one He^4 nucleus is produced (via reactions (6.2.6) and (6.2.2)) per He^3 nucleus destroyed as compared to one He^4 nucleus formed per two He^3 nuclei destroyed in the latter case. Thus the rate of energy generation at higher temperatures is greater than previously supposed by a factor two, neglecting small effects due to neutrino energy differences in the two different mechanisms for completing the proton-proton chain. We may examine the situation quantitatively as follows.

Let us write the number of reactions per second per cubic centimeter between nuclei of types j and k as

$$r_j = g_j n_j n_k \tag{6.3.4}$$

$$\text{or} \quad r_j = \frac{1}{2} g_j n_j^2, \tag{6.3.5}$$

where the latter expression refers to reactions between identical nuclei. The quantities g are the averages of cross section times velocity $\langle \sigma v \rangle$, discussed in Chapter 5. They are

---

*Editor's Note: Later published as [47].

related to the constants A and B of Table 6.1 by the expression

$$g_j = \frac{AA_j}{N_0} T^{-2/3} e^{B/T^{1/3}}, \tag{6.3.6}$$

where A_j is the mass number of the nucleus of type j, and N_0 is Avogadro's number. Then we may make the following definitions:

$$\text{for} \quad \text{H}^1(\text{p},\ \beta^+\nu)\text{H}^2(\text{p},\ \gamma)\text{He}^3,\ r_1 = \frac{1}{2} g_1 n_1^2;$$

$$\text{for} \quad \text{He}^3(\text{He}^3,\ 2\text{p})\text{He}^4,\ r_3 = \frac{1}{2} g_3 n_3^2;$$

$$\text{and for} \quad \text{He}^3(\alpha,\ \gamma)\text{Be}^7,\ r_4 = g_4 n_3 n_4,$$

where n_1, n_3, and n_4 are the number densities of H^1, He^3 and He^4, respectively.

For equilibrium, the rate of formation of He^3 must be equal to its rate of destruction:

$$\frac{1}{2} g_1 n_1^2 = \frac{1}{2} g_3 n_3^2 + g_4 n_3 n_4,$$

$$\therefore n_3 = \frac{g_4}{g_3} n_4 \left[\left(1 + \frac{g_1 g_3}{g_4^2} \frac{n_1^2}{n_4^2} \right)^{\frac{1}{2}} - 1 \right]. \tag{6.3.7}$$

It will turn out that for high temperatures,

$$\frac{g_1 g_3}{g_4^2} \frac{n_1^2}{n_4^2} \ll 1. \tag{6.3.8}$$

Then

$$n_3 \approx \frac{1}{2} \frac{g_1}{g_4} \frac{n_1^2}{n_4}. \tag{6.3.8}$$

This is equivalent to neglecting the $\text{He}^3(\text{He}^3, 2\text{p})\text{He}^4$ reaction. It will also turn out that for low temperatures:

$$\frac{g_1 g_3}{g_4^2} \frac{n_1^2}{n_4^2} \gg 1. \tag{6.3.9}$$

Then

$$n_3 \approx \left(\frac{g_1}{g_3} \right)^{1/2} n_1. \tag{6.3.9}$$

This is equivalent to neglecting the $\text{He}^3(\alpha, \gamma)\text{Be}^7$ reaction. The rates of destruction of

He^3 by the reactions He^3(He^3, 2p)He^4 and He^3(α, γ)Be^7 are equal when

$$\frac{1}{2}g_3 n_3^2 = g_4 n_3 n_4,$$

$$i.e. \quad \frac{g_1 g_3 n_1^2}{g_4^2 n_4^2} = 8. \tag{6.3.10}$$

From Table 6.1 and with the new values of S, we obtain:

$$g_1 = 5.5 \times 10^{-37} \; T^{-2/3} \; e^{-34.7/T^{1/3}}, \tag{6.3.11}$$

$$g_3 = 4 \times 10^{-12} \; T^{-2/3} \; e^{-122.5/T^{1/3}}, \tag{6.3.12}$$

and $\quad g_4 = 1.3 \times 10^{-15} \; T^{-2/3} \; e^{-128.0/T^{1/3}}. \tag{6.3.13}$

In equation (6.3.11) the small corrections to the proton-proton rate given in the expression on p. 53 have been neglected.

$$\therefore \quad \frac{g_1 g_3}{g_4} = \exp\left(-41.2 + \frac{98.8}{T^{1/3}}\right). \tag{6.3.14}$$

A newly-formed star may contain two-thirds or more of hydrogen by weight. Thus $n_1 \approx 8 n_4$. It follows from equation (6.3.14) for this case that the two modes of destruction of He^3 have equal rates at a temperature of 16.1×10^6 °K. The central regions of the sun when initially formed were at a temperature of about 14×10^6 °K, and hence the He^3(α, γ)Be^7 reaction was not important there.

A star of about a solar mass and about 5×10^9 years old has exhausted the majority of its central hydrogen. For such a star $n_1 \approx n_3$ at the center. For this case, the rates of the two He^3 reactions are equal at a temperature of 10.2×10^6 °K. Thus it is evident that the region of operation of the He^3(α, γ)Be^7 reaction in the sun has spread during the course of the sun's evolution. The extra rate of energy generation resulting from this new feature should be taken into account in calculations of stellar models.

If ϵ_1 is the rate of energy generation by the proton-proton chain under the usual assumption that the chain is completed by the He^3(He^3, 2p)He^4 reaction, then this rate must be multiplied by the following correction factor to get the correct rate ϵ_2 (but see

56

equation (6.3.24)):

$$\epsilon_2 = \epsilon_1 \left[\left(1 + \frac{g_1 g_3 n_1^2}{g_4^2 n_4^2} \right)^{\frac{1}{2}} + 3 \right] \Big/ \left[\left(1 + \frac{g_1 g_3 n_1^2}{g_4^2 n_4^2} \right)^{\frac{1}{2}} + 1 \right]. \qquad (6.3.15)$$

We must now consider the various mechanisms by which Be^7 can be destroyed in stellar interiors. In the laboratory Be^7 captures electrons from its K-orbit to decay with a half-life of 53 days to Li^7 (reaction (6.2.6)). In stars at temperatures in the vicinity of 15×10^6 °K, Be^7 is ionized most of the time. The half-life for K-capture would then be given by 106 days divided by the fraction of the time that Be^7 has one electron in its K-orbit. Fairly simple expressions for this probability can be obtained from the equations of dissociative equilibrium, but complications set in when one tries to take account of pressure ionization effects, and no expression for this rate will be given here.

In the interiors of stars on the main sequence, the density is of the order of 10^2 g/cm^3. All except heavy atoms are completely ionized, but the high density packs the electrons around the ions nearly as closely as the K-electrons in the neutral atoms. We may, therefore, expect that there will be a fairly appreciable probability for capturing these free electrons.

The rate for free electron capture may be estimated fairly crudely in the following way: We will use the rate which has been experimentally determined for capturing K-electrons, but we will replace the probability of finding a K-electron inside the nucleus by the probability of finding a free electron inside the nucleus, and we will multiply the single K-electron capture rate by the ratio of these probabilities. Convenient expressions for doing this have been given by Blatt & Weisskopf [32].

The probability of finding a K-electron inside the nucleus is

$$\int |\psi_e|^2 dV = \frac{1}{\pi} \left(\frac{Z m e^2}{\hbar^2} \right)^3 V_N, \qquad (6.3.16)$$

where ψ_e is the electron wave function, the integral is taken over the nuclear volume, m is the electron mass, and V_N is the volume of the nucleus. For free electrons in the absence of Coulomb fields, if an electron is confined to a volume V, the probability of finding it inside the nuclear volume is V_N/V. In the presence of a Coulomb field this must be multiplied by a Coulomb function:

$$\int |\psi_e|^2_{\text{free}} dV = \frac{V_N}{V} \frac{2\pi\eta}{1 - \exp(-2\pi\eta)}, \tag{6.3.17}$$

where

$$\eta = \frac{Ze^2}{\hbar v},$$

and where v is the velocity of the free electron.

We will identify the volume V with the volume occupied by one electron in the stellar interior. If the reasonable assumption is made that hydrogen contains one unit of atomic mass per electron and all other elements contain two units of mass per electron, then

$$V = \frac{3.32 \times 10^{-24}}{\rho(1 + x_1)} \text{ cm}^3, \tag{6.3.18}$$

where ρ is the density in g/cm^3 and x_1 is the concentration of hydrogen by weight.

For the small electron velocities which are involved here, η is sufficiently large that the exponential part of the denominator in (6.3.17) can be neglected compared to unity. The expression is then easily integrated over the Maxwell distribution of velocities, and the rate of free electron capture becomes

$$p_e = 1.9 \times 10^{-9} \frac{\rho(1 + x_1)}{T^{\frac{1}{2}}} \text{ s}^{-1}, \tag{6.3.19}$$

where the temperature T is in the usual units of 10^6 °K.

The expressions for the electron wave functions which have been used here are non-relativistic. The proper Dirac relativistic wave functions have a slightly different radial dependence than the nonrelativistic ones, but both the free and K-electron wave functions are changed in essentially the same way. Therefore there has not been much error introduced from this source. Nevertheless it is desirable to have a calculation done in which the proper Dirac wave functions for free electrons are introduced into the matrix element for the capture rate.

We now consider the destruction of Be^7 by thermonuclear reactions with hydrogen:

$$\text{Be}^7(\text{p}, \gamma)\text{B}^8, \tag{6.3.20}$$

followed by

$$B^8(\beta^+\nu)Be^8(2\alpha). \tag{6.3.21}$$

The ground state of B^8 is bound by 143 kev ([48]; T. Lauritsen, private communication). Therefore, the discussion of direct capture processes later on p. 67 applies here, since by analogy with the level structure of Li^8 it appears that the first excited state of B^8 will lie nearly 1 Mev above the ground state, and hence we are not near a resonance. A calculation of the rate of the reaction (6.3.20) proceeding by a direct nonresonant capture process has been carried out.

The cross section for capturing a proton into a featureless square well is [49]:

$$\sigma_{L,L\pm1} = \frac{256\pi^5e^2\nu^2}{3hc^3k^2v}(L + \frac{1}{2} \pm \frac{1}{2})\left|\int \bar{F}_L R_r^2 dr\right|^2, \tag{6.3.22}$$

where ν is the photon frequency, k and v are the electron wave number and velocity, respectively, $\bar{F}$ is the solution of the radial Schrödinger equation for the incoming proton wave, and R is the radial wave function of the final state normalized so that

$$\int_0^\infty R^2 r^2 dr = 1.$$

The ground state of B^8 is formed from that of Be^7 by adding a $p_{3/2}$ proton. Therefore, an incoming s-wave proton can transform to the desired p-wave proton by an electric dipole transition, and we take $L = 0$ in (6.3.22). The nucleus is represented by a square well. A relation between the depth of the potential and the radius of the well is determined by the requirement that the wave function of the final state and its logarithmic derivative must be continuous at the boundary of the well. We shall take the radius of the well to be a free parameter. Inside the well, the solutions of the radial wave equations are spherical Bessel functions; outside the well, the solutions must be obtained by numerical integration. For this purpose the W.K.B. approximation for the logarithmic derivative of the wave function of the final state given by Thomas [50] was used.

The integral in (6.3.22) was determined numerically for a proton energy of 20 kev in the center of mass system. The integrand changes sign near the boundary of the well, and the square of the matrix element evaluated over all configuration space is some

3×10^5 times larger than that part of it lying inside the nucleus. This illustrates the non-resonant character of the situation and indicates the danger involved in using dispersion theory far from resonances in cases of this sort.

As Wilkinson [51] has pointed out, equation (6.3.22) must be modified to take account of the facts: (1) we are only interested in those s to p transitions that lead to a final $1p_{3/2}$ state; (2) the effective proton charge is less than e; (3) the real reaction carries a different statistical weight factor from (6.3.22); (4) in the real system several equivalent $1p_{3/2}$ nucleons are involved so that care must be taken that initial and final state wave functions are properly antisymmetrical; and (5) the isotopic spin of the system in (6.3.22) is one-half, whereas that of the actual system can be zero or one. The corrections for these facts are dependent upon the nuclear coupling scheme which is used. Wilkinson [51] gives expressions for use with both LS and jj coupling. The correction factor so evaluated is 0.156 for LS coupling and 0.125 for jj coupling. In a proper intermediate coupling scheme, there is a chance that the correction factor can deviate considerably outside these limits, but this does not usually happen. Thus a final correction factor of 0.140 is adopted.

The results of the calculations, expressed in terms of the cross section parameter S, are

Well radius (cm)	S (ev barns)
3×10^{-13}	960
4×10^{-13}	1380
5×10^{-13}	1940

Since the result does not appear to depend critically upon the well radius parameter, and, since a well radius of 4.2×10^{-13} cm is a standard value, we will adopt $S = 1500$ ev barns as the result of this calculation.

Hence the proton capture rate becomes

$$p_p = 1.95 \times 10^{9-44.4/T^{1/3}} \frac{\rho x_1}{T^{2/3}} \ \mathrm{s}^{-1}. \tag{6.3.23}$$

In order to evaluate the importance of these destruction processes in the present sun, the writer has used the evolved solar model recently computed by R.L. Sears (private communication). In this model the central density is 180 g/cm^3, the central temperature is 16.1×10^6 °K, and the hydrogen concentration by weight has fallen to 0.42. It must, of course, be kept in mind that this model has been computed without taking into account the modifications to the proton-proton chain discussed here, or the carbon cycle which accounts for 30 percent of the energy generation in the model.

In this model, the $He^3(\alpha, \gamma)Be^7$ process completes the proton-proton chain within seven percent of the radius from the center, at which point $\rho = 135$ g/cm^3, $x_1 = 0.516$, and $T = 14.9 \times 10^6$ °K. This conclusion is a conservative one because electron shielding corrections have not been applied to the cross section factor S for the $He^3(\alpha, \gamma)Be^7$ process. Within this region about 12 percent of the energy generation by the p-p chain takes place. Also, within this region p_p exceeds p_e by factors varying from 2 to 1. Thus, it appears likely that the proton-proton chain is completed primarily by the formation of B^8 at the center of the sun.

The decay of B^8 has some particularly interesting properties. The beta-decay transition is an allowed one which proceeds through the broad 3 Mev excited state of Be^8, giving a positron endpoint energy of about 14 Mev. The neutrinos emitted with the positrons have the same endpoint energy and an average energy of about 7 Mev. In this case 26 percent of the energy in the conversion of hydrogen to helium is carried away from the sun by the B^8 neutrinos. Thus the rate of effective energy generation in the center of the sun is not given by equation (6.3.15), but by

$$\epsilon_3 = \left(\frac{p_e + 0.74 p_p}{p_e + p_p} \right) \epsilon_2. \tag{6.3.24}$$

It should be noted that there is a faster rate of hydrogen depletion at the center of the sun associated with this energy generation rate than is normally the case.

The flux of B^8 neutrinos to be expected at the earth is about 4×10^9 neutrinos/cm^2 s. These neutrinos are very energetic and quite effective in inducing inverse beta reactions. In particular, R. Davis (private communication) estimates that he can detect this flux of neutrinos by placing 5,000 or more gallons of carbon tetrachloride in a deep mine where cosmic ray effects are absent. The reaction induced in this case is $Cl^{37}(\nu, e^-)A^{37}$.[†]

[†]Editor's Note: The Commission on Inorganic Nomenclature adopted a change of the chemical symbol of Argon from 'A' to 'Ar' at the Conference of the International Union in 1957 [52].

Later in Section 18.6, we will mention giant stars with unusually large abundances of lithium. Recent work suggests that strong lithium lines exist in a fairly large percentage of the giant stars. It was suggested that magnetic activity might produce this abundance anomaly by spallation reactions, but the present considerations seem to allow a more satisfactory theory to be put forth which was suggested by the writer some time ago [53].

It is known that giant stars have deep outer convection zones, but the depth to which these zones go is extremely uncertain. The process of convection usually involves mass motions in a regular cellular pattern, but the size of the cells in stellar atmospheres is unknown. Transport times within the cells may only be a few hours, however. Consider convection zones which go down to temperatures of the order of 10×10^6 °K. If considerable amounts of the hydrogen have been converted to helium, then there will be a great deal of Be^7 formed in such temperature regions. The Be^7 may then be transported rapidly to regions of lower temperature at which Li^7 has a fairly long life against proton capture. Mixing may then introduce some of the Li^7 into other convective cells, and some of it can reach the surface in what is essentially a diffusion process with a long mean free path. A diffusion equilibrium will be set up in which the amount of Li^7 coming to the surface is equal to that diffusing to destruction in the interior. One may expect there to be a correlation between weaker hydrogen lines and stronger lithium lines in these lithium stars, if these ideas are correct. Detailed analysis of the conditions involved may be an excellent tool for research in the understanding of stellar atmospheric convection.

It is also of interest to speculate whether the presence in the sun of the $He^3(\alpha, \gamma)Be^7$ reaction may be a contributing cause of the ice ages on the earth. Öpik [54; 55; 56][‡] has suggested that regions in the sun can be set up which are metastable against convection. If a disturbance causes a convective mixing to occur, the addition of material richer in hydrogen nearer the solar center will cause a temperature fluctuation in the sun which will require millions of years to damp out. When the $He^3(\alpha, \gamma)Be^7$ reaction sets in, there is an increase in the rate of energy production due to more efficient conversion of hydrogen to helium, and it may be profitable to investigate whether this can cause such metastable layers to be created.

[‡]Editor's Note: Results from the latter reference were later refereed and published as [57].

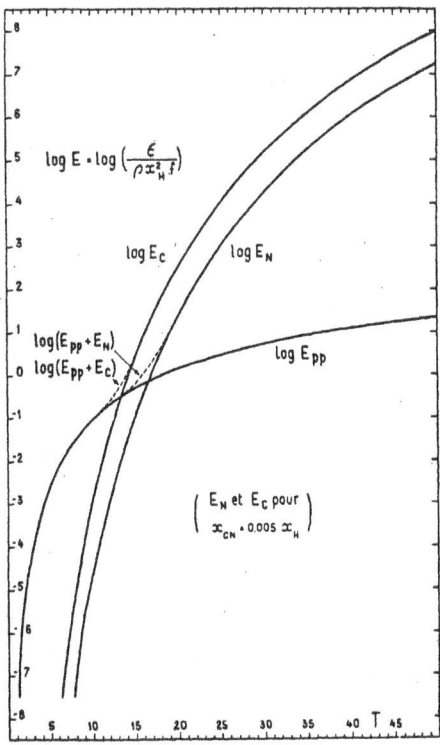

Figure 6.4.1: The rates of energy generation from the proton-proton chain and from the carbon cycle. The carbon-cycle values are computed for the assumptions that the $C^{12}(p, \gamma)N^{13}$ and that the $N^{14}(p, \gamma)O^{15}$ reactions are the slowest in the cycle, using the nuclear cross sections at low energies currently in the literature [58]. The energy generation rates are in ergs/g s, and the temperature is in units of 10^6 °K.

6.4 The carbon-nitrogen cycle

The hotter stars of the main sequence, with central temperatures of 17×10^6 °K and higher, convert hydrogen to helium mainly by the carbon-nitrogen cycle. The reactions in this cycle are:

$$C^{12}(p, \gamma)N^{13}, \tag{6.4.1}$$

$$N^{13}(\beta^+\nu)C^{13}, \tag{6.4.2}$$

$$C^{13}(p, \gamma)N^{14}, \tag{6.4.3}$$

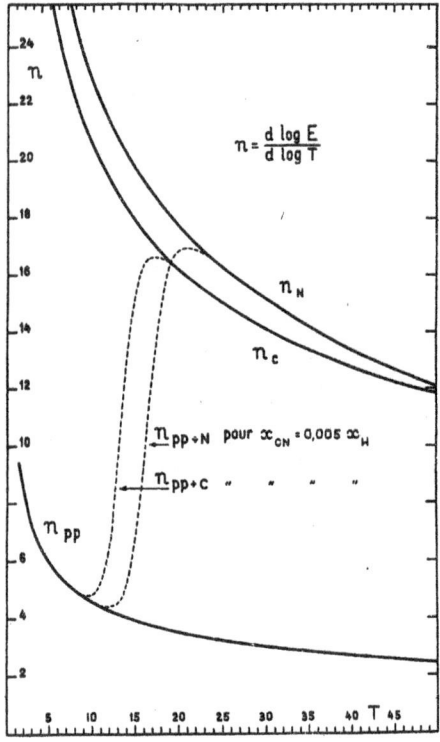

Figure 6.4.2: The power-law exponent in the temperature sensitivity equation $p = p_0 T^n$, for the proton-proton chain, the carbon cycle, and the combination of both. The carbon cycle is alternately assumed to have the $C^{12}(p, \gamma)N^{13}$ and $N^{14}(p, \gamma)O^{15}$ reactions slowest [58]. The temperature is in units of 10^6 °K.

$$N^{14}(p, \gamma)O^{15}, \tag{6.4.4}$$

$$O^{15}(\beta^+ \nu)N^{15}, \tag{6.4.5}$$

$$\text{and} \quad N^{15}(p, \alpha)C^{12}. \tag{6.4.6}$$

Here, the carbon and nitrogen nuclei act as catalysts for the conversion of protons into alpha-particles. We must also consider the reaction

$$N^{15}(p, \gamma)O^{16}, \tag{6.4.7}$$

which removes nuclei from the cycle. Bethe [59] estimated on crude grounds that (6.4.7) was only 10^{-4} times as likely as (6.4.6). The writer currently estimates that the ratio of the two reaction rates accidentally turns out to be very close to 1.0×10^4. According to Goldberg, Müller & Aller [60], there are 2200 hydrogen atoms per carbon or nitrogen atom in the sun. Hence the carbon cycle would have to turn over 550 times to exhaust the hydrogen in the sun (actually, it is not important in the sun), and it would have to turn over about 10^4 times in a Population II star in which the C, N group was only 5 percent of the value in the sun.[§] We see that in the latter case we must consider leakage of carbon-cycle nuclei to O^{16} via reaction (6.4.7).

The carbon-cycle reaction rates given in Table 6.1 are based on measurements made at the Kellogg Radiation Laboratory of the California Institute of Technology. However, the value for the $N^{14}(p, \gamma)O^{15}$ reaction is not the one available in the literature, since the original low energy measurement for this reaction appears to be in error (W. A. Fowler, private communication). The actual reaction rate appears to be lower than that in the literature by about an order of magnitude (W. A. S., Lamb & R. E. Hester, private communication). Hence the value in Table 6.1 has been reduced by a factor 10 from the value given by Fowler [41]. This reaction cross section is currently being remeasured using the 100 kev injector to the materials testing linear accelerator at Livermore, which can deliver a beam current of nearly one ampere. The final results will be of real interest. Figures 6.4.1 and 6.4.2 show the C-N reaction rates and power-law exponents compared to those of the p-p chain, using the old values in the literature.

The equilibrium abundances of the carbon and nitrogen isotopes at various temperatures are of considerable interest because some of them can be checked astrophysically in sources in which the gas has been subjected to high temperatures. The most abundant isotope is N^{14}. The other isotopes have abundances relative to N^{14} which are inversely proportional to the respective reaction rates. The carbon isotopes have much greater abundances at high temperatures than at low temperatures. N^{15} always has a very low abundance owing to its large reaction rate, which comes from a high nuclear cross section associated with particle emission. The ratio of the C^{12} to C^{13} abundances is always close

---

[§]Editor's Note: Part of this sentence and the next one are missing in the 2nd edition, likely owing to a copying omission of a line at the bottom of a page.

to 4. This ratio is often observed in stars with N spectra.

It should be noted that the relative numbers of carbon-cycle nuclei in the sun does not correspond to carbon-cycle equilibrium proportions. The ratio of C^{13} to N^{14} is about what would be expected for the carbon cycle operating in the range of 20 to 30×10^6 °K, but there is a big excess of C^{12} and of N^{15}.

6.4.1 Supplementary Notes: The $N^{14}(p, \gamma)O^{15}$ reaction

Lamb & Hester [61] have determined the cross section to be represented by $S = 2700 \pm 200$ ev barns. This confirms the lower values discussed in the text.

6.5 Oxygen and fluorine reactions

Hydrogen thermonuclear reactions involving nuclei heavier than mass number 15 are usually slower than the carbon-cycle reactions. However, since we are interested in conditions in which the carbon cycle turns over many times, we must consider whether any astrophysically significant changes will be produced in the abundances of heavier nuclei during the period when hydrogen is being consumed. As we shall see, conditions sometimes arise in which hydrogen is consumed at temperatures as high as 50 or 60×10^6 °K. These higher temperatures favor some thermonuclear reactions with heavier nuclei.

The nuclear reactions of interest which involve oxygen and fluorine nuclei are:

$$O^{16}(p, \gamma)F^{17}, \tag{6.5.1}$$

$$\text{followed by} \quad F^{17}(\beta^+\nu)O^{17}, \tag{6.5.2}$$

$$O^{17}(p, \alpha)N^{14}, \tag{6.5.3}$$

$$O^{18}(p, \alpha)N^{15}, \tag{6.5.4}$$

$$\text{and} \quad F^{19}(p, \alpha)O^{16}. \tag{6.5.5}$$

We see that these reactions feed nuclei into the carbon cycle. They are therefore of interest in that they may make the more abundant O^{16} nuclei available as catalysts in the carbon-nitrogen cycle, particularly in Population II stars in which there can be significant leakage of C-N nuclei to O^{16} via reaction (6.4.7).

All of the above reactions except (6.5.1) take place quite quickly because they involve exothermic particle emission. Salpeter [8] has estimated the reaction rate for (6.5.1), but his value is believed to be a lower limit for the following reasons.

Warren, Laurie, James, & Erdman [62] have studied the $O^{16}(p, \gamma)F^{17}$ reaction and have found that it is nonresonant. Most of the radiation goes to a 0.5 Mev excited state in F^{17} and only 10 percent to the ground state of F^{17}, for bombarding energies in the range 0.8 to 2.1 Mev. The calculation of Salpeter is based on the assumption that reaction (6.5.1) takes place by p-wave proton capture through a very broad state at about 4 Mev excitation energy in F^{17}, and the reaction cross section has been extrapolated to thermonuclear bombarding energies using the single-level Breit-Wigner formula and taking into account the variation in the radiation width with excitation energy in F^{17}. Only ground state transitions were assumed. One particularly striking feature of the 0.5 Mev level in F^{17} is that it lies just barely below the sum of the masses of O^{16} and a proton (it is bound by about 80 kev), and it has a reduced proton width which is of the order of the single particle limit. The latter fact indicates that a good representation of the wave function of the state can be obtained by assuming that a proton moves in the potential well due to the O^{16} nucleus, and the former fact indicates that this wave function has a very large extension in configuration space, very much larger than the extension usually considered when setting up internal and external regions in nuclear dispersion theory. When a proton is incident on the O^{16} nucleus and captured, most of the captures take place in the external region. This is a direct, non-resonant process to which the Breit-Wigner formula does not apply. Some very crude calculations of the writer indicate that the nuclear cross section at low energies is considerably larger than that due to Salpeter's resonant extrapolation. Hence, Salpeter's value of the reaction parameter S is taken to be a lower limit in Table 6.1.

The relative reaction rates of $N^{14}(p, \gamma)O^{15}$ and $O^{16}(p, \gamma)F^{17}$ reactions between temperatures of 20 and 60×10^6 °K are shown in Table 6.2. It may be seen that the $O^{16}(p, \gamma)F^{17}$ reaction is at least as fast as 10^{-4} to 10^{-3} times the $N^{14}(p, \gamma)O^{15}$ reaction. Hence, the oxygen in Population II stars is utilized in carbon-cycle reactions, and it is very likely that it is also utilized in stars of solar composition.

This gives one possible explanation of stars with R and M spectra. These are red giant stars in which the molecular compounds which characterize the spectra are carbides rather than oxides. In cool atmospheres, one of the most abundant molecules, because of its large binding energy, is CO (which does not give bands in the observable region of the spectrum). Nearly all the carbon or oxygen, whichever is of lesser abundance, is in the form of CO. The atom of larger abundance is then available to form compounds with other kinds of atoms. Oxygen is normally the more abundant atom; hence oxide formation occurs in normal stellar atmospheres. However, if the material in a stellar atmosphere is passed through a region of high temperature for an extended period of time, then much of the oxygen may be converted to carbon and nitrogen, leaving carbon more abundant than oxygen. However, in order for this explanation to work, it is necessary for nearly all the oxygen to be converted and for the characteristic temperature to be very high, since N^{14} has a much greater equilibrium abundance in the carbon cycle than C^{12}. It is also necessary that the C^{12} to C^{13} ratio be close to 4; this condition is satisfied by most but not all of the carbon stars.

6.5.1 Supplementary Notes: The $O^{16}(p, \gamma)F^{17}$ reaction

Low-energy cross sections have been measured for this reaction at the California Institute of Technology, at Livermore, and at the University of British Columbia. These measurements are all consistent with the value $S = 6000 \pm 2000$ ev barns. This is 60 times larger than the lower limit given in Table 6.1. We may thus see from Table 6.2 that the oxygen in a star running on the carbon cycle will be converted into carbon-cycle isotopes with a $1/e$ depletion rate comparable to about the time required to convert 10 percent of the original material from hydrogen to helium. This is important for stellar model calculations because there is about three times as much oxygen in stars like the sun as carbon and nitrogen combined.

It is interesting to note that calculations of direct proton capture by the method outlined in Section 6.3.2 have been done at the California Institute of Technology. The result agrees with the experimental value very well.

6.6 The neon-sodium cycle

Hydrogen thermonuclear reactions with the neon and sodium isotopes have been discussed by Salpeter [8], Fowler, Burbidge, & Burbidge [63], and Marion & Fowler [42]. The reactions which may be of importance are:

$$Ne^{20}(p, \gamma)Na^{21}, \tag{6.6.1}$$

$$\text{followed by} \quad Na^{21}(\beta^+\nu)Ne^{21}, \tag{6.6.2}$$

$$Ne^{21}(p, \gamma)Na^{22}, \tag{6.6.3}$$

$$\text{followed by} \quad Na^{22}(\beta^+\nu)Ne^{22}, \tag{6.6.4}$$

$$Ne^{22}(p, \gamma)Na^{23}, \tag{6.6.5}$$

$$\text{and} \quad Na^{23}(p, \alpha)Ne^{20}, \tag{6.6.6}$$

$$\text{or} \quad Na^{23}(p, \gamma)Mg^{24}. \tag{6.6.7}$$

The reaction rate constants for these reactions have been taken from the discussion of Marion & Fowler, except that their value of the $Ne^{21}(p, \gamma)Na^{22}$ reaction rate is taken to be a lower limit.

The $Ne^{20}(p, \gamma)Na^{21}$ reaction proceeds through the tail of a resonance at the negative bombarding energy of -26 kev. Marion & Fowler have determined the reduced proton width of this level by a very elegant method which uses nuclear dispersion theory to interpret the difference in the excitation energies of this level and its mirror level in Ne^{21}. The biggest uncertainty in the reaction rate of (6.6.1) is the value assumed for the radiation width of the capturing level. Marion & Fowler used an average radiation width for magnetic dipole transitions in light nuclei [64], but Wilkinson's values have a spread of an average factor of 20. The error in the $Ne^{20}(p, \gamma)Na^{21}$ reaction rate may therefore also be estimated to be a factor 20.

The reaction rates for the $Ne^{20}(p, \gamma)Na^{21}$ reaction are also compared with those of the $N^{14}(p, \gamma)O^{15}$ reaction in Table 6.2. It may be seen that no appreciable conversion of Ne^{20} to Ne^{21} takes place even in fairly extreme Population II stars unless the magnetic dipole

Table 6.2: Values of $p/\rho x_1$

Temperature (10^6 °K)	$N^{14}(p,\gamma)O^{15}$	$O^{16}(p,\gamma)F^{17}$	$Ne^{20}(p,\gamma)Na^{21}$
20	1.7×10^{-16}	$> 4.4 \times 10^{-20}$	5.4×10^{-22}
30	2.3×10^{-13}	$> 7.8 \times 10^{-17}$	2.1×10^{-18}
40	1.7×10^{-11}	$> 9 \times 10^{-15}$	4.7×10^{-16}
50	3.5×10^{-10}	$> 2.3 \times 10^{-13}$	1.7×10^{-14}
60	3.5×10^{-9}	$> 3.2 \times 10^{-12}$	2.7×10^{-13}

radiation width is much higher than Wilkinson's average value. Thus it is completely uncertain at present whether any appreciable conversion of Ne^{20} to Ne^{21} will take place in any of the stars before hydrogen is exhausted.

Marion & Fowler have estimated that the $Ne^{21}(p,\gamma)Na^{22}$ reaction rate is comparable to their value for $Ne^{20}(p,\gamma)Na^{21}$. This relies on the fact that Broström, Huus & Koch [65] observed only one definite level in the range of bombarding energy 0.6 to 1.3 Mev. Hence, Marion & Fowler assumed that the level nearest the Gamow peak is about 400 kev away. However, Broström *et al.* observed evidence for many more weaker levels in the $Ne^{21}(p,\gamma)Na^{22}$ reaction, and so the level distance is probably much less than assumed by Marion & Fowler. T. D. Newton's [66] level density formula, although of questionable validity for such a light nucleus, predicts that 6 levels, formable by s-, p-, and d-wave protons, will lie in a 50 kev interval about the Gamow peak. Hence it is not unlikely that the $Ne^{21}(p,\gamma)Na^{22}$ reaction may be thermally resonant, and hence that its reaction rate is enormously greater than the values given in Table 6.1.

The other reactions in the neon-sodium cycle are relatively fast. It appears likely that the $Na^{23}(p,\alpha)Ne^{20}$ reaction rate is much faster than the $Na^{23}(p,\gamma)Mg^{24}$ rate. Hence there would be only a small rate of leakage of nuclei, from a hypothetical neon-sodium cycle.

To summarize, it is a completely open question as to whether the $Ne^{20}(p,\gamma)Na^{21}$ reaction can take place to an appreciable extent in any of the stars; on the basis of average nuclear parameters, it would not. If it does, it seems likely to be the slowest reaction in the neon-sodium cycle, so that any Ne^{21} formed is likely to be fairly quickly returned to Ne^{20} by further reactions. However, this second conclusion is by no means certain either.

6.6.1 Supplementary Notes: The $Ne^{20}(p, \gamma)Na^{21}$ reaction

The measurement by Pixley, Hester & Lamb [67] gives $S = 6 \times 10^4$ ev barns. This is five times the value given in Table 6.1 and indicates that the Ne^{20} can be quite extensively converted into Ne^{21} in stars with high temperatures in shell sources. The question of whether the Ne^{21} is then destroyed thus becomes most important in considerations of the possibility that neutrons will be produced by the $Ne^{21}(\alpha, n)Mg^{24}$ reaction.

Chapter 7

Stellar Evolution

We are now ready to consider the results of detailed computations of evolutionary sequences of stellar models. So far these sequences have become available only for the hydrogen consumption stages of stars not much more massive than the sun. Later, when we are interested in the more advanced stages of stellar evolution, we will have no detailed stellar models to guide us, and we will have to take a general approach when discussing the likely course of nuclear reactions in stellar interiors.

7.1   The gravitational contraction phase

The tracks followed in the Hertzsprung-Russell diagram by stars have been obtained with the aid of an electronic computer by L.O. Henyey [68] with collaborators. It was found that the stars follow a nearly-horizontal course to the left in the H-R diagram until they reach the vicinity of the main sequence. At this point, the luminosity of the stars decreases slightly, and the stars settle onto the main sequence for an extended period while they exhaust their central hydrogen. Only the more massive stars were considered, those which operate on the carbon cycle and have central convective zones surrounded by radiative envelopes. The times spent on the main sequence by these stars have been obtained, but detailed results have not yet been published.

Henyey and his collaborators have not taken into account the presence of extensive outer convection zones in their gravitationally contracting models, nor have they considered the effects of deuterium consumption. Now it is generally found that outer convection zones cause a decrease in luminosity of faint main sequence stars, so it is possible that the gravitationally contracting tracks in the H-R diagram may be lowered on this account.

It has been found that the abundance of lithium in the sun is only about one percent of the abundance in the earth relative to elements like sodium [69]. On the other hand, the abundance of beryllium is essentially normal [70]. This indicates that the material in the solar surface has been raised to a temperature in the range 3 to 4×10^6 °K at some period, sufficient to destroy most of the lithium but very little of the beryllium. The recent solar model of Schwarzschild, Howard, & Härm [71] contains an outer convection zone which extends inward only to a temperature close to 1×10^6 °K, which will destroy only deuterium. Hence it is evident that the solar convection zone must have been sufficiently more extensive in the gravitation contraction stage to reach temperatures of about 3.5×10^6 °K and to reach them at a time when the central temperature of the sun was not a great deal higher than this figure. Hence the sun may have been completely convective during the even earlier deuterium consumption stage.

7.2   Evolution of the sun

Schwarzschild, Howard, & Härm [71] have found from computations that the sun had a central temperature of 13×10^6 °K and a central density of 90 g/cm^3 when it first settled onto the main sequence. It then had a radiative core and envelope surrounded by an outer convection zone. It may be noticed from Figures 6.4.1 and 6.4.2 that, at 13×10^6 °K, the energy generation is almost entirely by the proton-proton chain, and the generation varies only as the fourth power of the temperature. This does not cause a very high temperature gradient at the center of the sun, and hence the centers of the sun and fainter stars are in radiative equilibrium. Schwarzschild *et al.* followed the change in the solar model which occurs as hydrogen is converted into helium. Originally 80 percent by weight of the material in the sun was hydrogen. Now the hydrogen abundance of the center of the sun is only 30 percent by weight. In order to maintain the energy generation the central temperature has risen to 15×10^6 °K and the central density to 130 g/cm^3.

The outer convection zone in the sun occupies 18 percent of the radius but contains only 0.3 percent of the mass. The sun has become brighter by half an astronomical magnitude during the last 5×10^9 years. This means that 2×10^9 years ago the solar luminosity was 20 percent less than now, and, if the surface temperature of the earth varies as the fourth root of the solar luminosity, the average terrestrial surface temperature was then equal to the freezing point of water. This may have had fairly significant effects on the evolution of life on the earth.

7.2.1 Supplementary Notes: Solar model

Some results of the computer calculation by R.L. Sears of evolved solar models have been given in Section 7.2.

7.3 Evolution of Population II globular cluster stars

The evolutionary tracks of globular cluster stars after they leave the main sequence have been shown in Figure 3.3.1. It may be seen that there is a gradual rise off the main sequence, a slight swing to the right, a rapid rise in luminosity at constant surface temperature, a swing upwards and to the right until the tip of the red giant branch is reached, followed by a decrease in luminosity and a movement to the left of the diagram on the horizontal branch. The principle features of these evolutionary sequences have been explained in a series of models computed by Hoyle & Schwarzschild [72]. More accurate results from an electronic computer have been obtained by Haselgrove & Hoyle, but their results have not yet appeared in print*. They have been briefly discussed by Bondi [74].

Haselgrove & Hoyle assumed a mass of 1.26 times that of the sun for the typical globular cluster star whose evolution is to be followed, since the resulting sequence of models compares well with the appearance of the H-R diagram for the cluster M3. After the star settles onto the main sequence, the hydrogen is consumed in the central regions by the proton-proton chain. As in the case of the sun, as hydrogen becomes depleted, the central temperature and density increase. After some time the carbon cycle takes

---

*Editor's Note: Published as [73].

over the burden of energy generation, inducing a convection zone at the center of the star which includes about eight percent of the mass. This convection zone exhausts the central hydrogen after a period of some 5×10^9 years. During this time the luminosity of the star has increased about 0.75 astronomical magnitudes without appreciable change in surface temperature. The corresponding track lies straight up in the H-R diagram.

After the central hydrogen has been exhausted, a change in the structure of the star must take place. The energy generation must now come from the hydrogen surrounding the helium core. Because of the strong dependence of the energy production on temperature, the effective thickness of the energy-producing region is only a small fraction of that of the core. Hence this energy-producing zone will be called a shell source of energy. Initially, there is a temperature gradient within the core. This cannot persist very long because the only source of energy in the core is now gravitational contraction, and at this stage of evolution practically no energy is released in the core by gravitational contraction. Hence the core must become isothermal. Haselgrove & Hoyle find that the stellar luminosity suddenly increases by half of a magnitude when the transition to the shell source takes place. They suggest that in a more massive Population I star, this transition to a shell source may cause the star to jump across the Hertzsprung gap.

As the shell source generates energy, it consumes more hydrogen, and the mass of the inert helium core increases at a rate proportional to the luminosity of the star. The luminosity of the star continues to increase. The actual track in the H-R diagram goes partly to the right as well as up, indicating that the stellar envelope is expanding slightly during this time. Haselgrove & Hoyle find that it takes an additional 1×10^9 years for the star to reach the point at which the track in the H-R diagram turns sharply upward.

At this point an extensive outer convection zone sets in, and the star increases rapidly in luminosity. The presence of an appreciable abundance of metals in the star would cause a considerably different track to be followed at this point, one lying much to the right of that of the globular cluster stars. However, this is almost entirely due to conditions of greater opacity in the envelope, and there would be little effect on central conditions.

The shell source continues to operate as the star progresses all the way up to the tip of the giant branch. As the evolution becomes advanced, the temperature in the shell source increases, until near the tip of the giant branch temperatures of 50 or 60×10^6 °K can be expected, taking account of the new low reaction rate for the $N^{14}(p, \gamma)O^{15}$ reaction in Table 6.1. At the tip of the giant branch, about half of the mass of the star is in the inert helium core. Haselgrove & Hoyle find that a little more than 6.5×10^9 years is needed for the stars to reach the tip of the giant sequence in M3.

At this stage the core is no longer isothermal. The very high stellar luminosity indicates that material is being rapidly added to the core. The core contracts quite rapidly, and there is now a fairly rapid release of gravitational energy in it. At the tip of the giant sequence the central density has reached about 10^5 g/cm^3, and the central temperature has reached the vicinity of 100×10^6 °K. The center of the star is a degenerate gas, and energy transport is by conduction, a very rapid process which keeps the temperature gradient very small at the center of the star. The shell source lies in a nondegenerate region, and most of the temperature difference between it and the center occurs in the outermost parts of the core.

This sequence of events is radically changed at the tip of the giant branch by the on-set of helium thermonuclear reactions. We have already seen that a degenerate gas cools when it contracts and heats when it expands. Now, in an ordinary star, there is a natural balancing process; if thermonuclear reactions start generating too much energy, the star expands, the gas cools, and the thermonuclear reactions are quenched. This will not hap-pen when helium thermonuclear reactions start in our degenerate core. Once the central temperature has risen to the point where the energy generation exceeds the outward flux of energy, the expansion that occurs is accompanied by heating and by an increasing excess of energy production. Hence the situation is very unstable, and at some point, a rapid expansion of the core, accompanied by a great deal of energy production, must take place. This will continue until the core material has become nondegenerate and the natural balancing process can quench the central reactions. The energy generation will take place throughout most of the core because the high conductivity of the degenerate material maintains the temperature nearly constant throughout the expansion until the material becomes nondegenerate, and its opacity increases.

Current calculations with stellar models do not take us beyond the onset of this first central instability in a star. Therefore any further statements about stellar evolution must be very speculative. We must rely on the empirical evidence of Figure 3.3.1 that our stellar models should next occupy the horizontal branch and must eventually wind up as white dwarf stars. It is quite likely that some of the stars will attain very high central temperatures and densities in this process. The situation is complicated because it has also been observed that some stars eject mass into space at these advanced stages of evolution.

Hoyle & Schwarzschild [72] calculated some preliminary stellar models in which it was assumed that the energy generation came from central helium reactions and a hydrogen shell source. These models lie in the region of the horizontal branch but somewhat above it. Furthermore, they assume that the distribution of chemical composition in the star was not altered following the expansion of the core. This is questionable.

During the core expansion, it is likely that a great deal of energy will be generated in the central regions. After the matter has become nondegenerate, there will be a very steep temperature gradient in the central regions which must drive an extensive and vigorous convection zone for a short period of time. This convection zone is likely to extend into the radiative hydrogen envelope and to mix a lot of hydrogen into the interior regions.

One possibility is that the star will be nearly uniformly mixed throughout (M. Schwarzschild, private communication). In this case it may jump to the left of the horizontal branch and then work its way up to the giant branch again as the central hydrogen is exhausted once more. If the mixing is much less extensive than this, then it is possible that hydrogen would not succeed in mixing all the way to the center, owing to the high rate of energy generation from the carbon cycle and local expansion of the volume elements containing the hydrogen when they enter regions of high temperature. It is also possible that the hydrogen is mixed to the center but very rapidly exhausted there at temperatures in excess of 100×10^6 °K. These different possibilities will be of considerable interest to us when we consider heavy element synthesis by neutron production and capture.

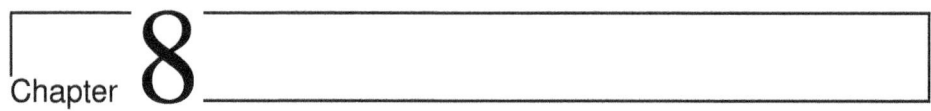

Chapter 8

Helium Thermonuclear Reactions

He^4 is stable in a stellar interior until very high temperatures are reached. The reason for this is that the Li^5 and Be^8 nuclei are unstable against particle emission; hence a He^4 nucleus cannot capture a proton or another alpha-particle except momentarily. However, when we go to very high temperatures and densities, the protons have disappeared but the momentary captures of one alpha-particle by another assume great importance.

8.1 The formation of carbon

Be^8 is unstable to a break-up into alpha-particles by 94 ± 1 kev (W. A. Fowler, private communication*). Hence at high temperatures and densities when $kT \simeq 10$ kev, small amounts of Be^8 can be maintained in thermal equilibrium with the helium. These amounts can be calculated from statistical mechanics; we need to know only the spin and disintegration energy of Be^8, and that the conditions for statistical equilibrium are satisfied. These conditions will be discussed later. The most important condition is that there should be reactions which form Be^8 in a time short compared to the lifetime of the stage of stellar evolution in which we are interested, and that the Be^8 half-life should also be short compared to that time. These conditions are well satisfied when the temperature reaches 100×10^6 °K and the density reaches 10^6 g/cm^3. The collision rate between the alpha-particles is then extremely high, and the Be^8 half-life is of the order

---

*Editor's Note: Later published [75].

of 10^{-15} s [76].

Hoyle [77] has written down the number density of Be^8 nuclei from statistical mechanics. It is:

$$n_2 = 9.40 \times 10^{-31-473/T} \frac{n_1^2}{T^{3/2}}, \qquad (8.1.1)$$

where n_1 is the number of He^4 nuclei per cm^3 and n_2 is the number of Be^8 nuclei per cm^3. Nakagawa, Ohmura, Takebe, & Obi [78] have shown that the previous relation can be calculated using the same form of development that culminated in equation (5.2.2); their derivation has the further useful property of showing that the Be^8 nuclei formed have a Maxwell distribution of velocities. It is important to be sure of this point because the Be^8 nuclei decay before they have time to make many collisions with the other nuclei present. However, Nakagawa *et al.* have neglected to include the statistical factor of $1/2$ which must be used when we deal with collisions of like nuclei (this expresses the fact that whereas there are $N_1 N_2$ pairs of unlike nuclei in a gas, there are only $N^2/2$ pairs of like nuclei).

The Be^8 nuclei can capture an additional alpha-particle:

$$Be^8(\alpha, \gamma)C^{12}. \qquad (8.1.2)$$

At $T = 100 \times 10^6$ °K and $\rho = 10^5$ g/cm^3, there are only about one part in 10^9 of the He^4 nuclei in the form of Be^8. However, the capture of alpha-particles is a resonant reaction which takes place through the second excited state of C^{12} [79], whose excitation energy is 7.654 ± 0.003 Mev (W. A. Fowler, private communication[†]). The reaction rate of the $Be^8(\alpha, \gamma)C^{12}$ reaction is given by equation (5.2.2), in which it is necessary to know certain nuclear parameters.

Fowler, Cook, Lauritsen, Lauritsen & Mozer [80] have shown in a very nice experiment that the second excited state of C^{12} can be formed by alpha-particles. They observed the beta decay of B^{12} to this state; it was followed by alpha-particle emission. The energies of alpha-particles were measured very accurately, giving the Be^8 disintegration energy and the C^{12} excitation energy listed in this section. The spin and parity of the 7.654 Mev state are 0^+ (the experimental evidence for this assignment is reviewed in

---

[†]Editor's Note: Later published [75].

a forth-coming paper by W.A. Fowler[‡]). The alpha-particle width of the state is more than 100 times the radiation width (W. A. Fowler and H. E. Gove and A. E. Litherland, private communication). Hence only the radiation width needs to be known to give the reaction rate in equation (5.2.2). R.A. Ferrell (private communication to W. A. Fowler) has computed the radiation width for electric quadrupole radiation to the first excited state on the basis of a nuclear model which has been quite successful in predicting other properties in C^{12}. His result is that $\Gamma_\gamma = 1.38 \times 10^{-3}$ electron volts with an estimated error of a factor of two. This error is the main source of error in the rate for reaction (8.1.2). Salpeter has pointed out (E. E. Salpeter, to be published) that one must include a statistical factor of one-third in the reaction rate. This is a result of the fact that we really have three identical particles colliding to form C^{12}, and so we must take the number of combinations of like particles taken three at a time. This number is $N^3/6$, but we have already included part of this in taking $N^2/2$ as the number of pairs available of to form Be^8.

Therefore the number of reactions per second per alpha-particle which form C^{12} is:

$$p_1 = 4.4 \times 10^{-48 - 1878/T} \frac{n_1^2}{T^3} \text{ s}^{-1}, \qquad (8.1.3)$$

$$= q_1 n_1^2, \qquad (8.1.4)$$

$$\text{where } q_1 = \frac{4.4 \times 10^{-48-1878/T}}{T^3}. \qquad (8.1.5)$$

8.1.1 The triple alpha reaction

Salpeter [81] has pointed out that a better way of calculating the rate of the $Be^8(\alpha, \gamma)C^{12}$ reaction is to compute the equilibrium concentration of C^{12*}, the second excited state of C^{12} at 7.654 Mev, in the helium gas under conditions of statistical equilibrium. This holds to a good approximation because the alpha-particle width of the state is much larger than the radiation width. The results of this calculation have been incorporated in the text; the change is a rather small one. It should be pointed out that the radiation width has still been assumed to be 0.00138 electron volts. It is disappointing that the radiative decay of this state has not yet been detected experimentally.

[‡]Editor's Note: Published as [75].

8.2   The formation of oxygen

The carbon nuclei which have been formed can capture further alpha-particles to form oxygen:

$$C^{12}(\alpha, \gamma)O^{16}. \tag{8.2.1}$$

There are no resonances in the thermonuclear region for this reaction, and hence the nonresonant contributions from rather distant levels must be taken into account. A survey of the levels in the O^{16} nucleus indicates that the following levels probably give the main contribution to the thermonuclear reaction rate. The thermonuclear energy region lies at an excitation energy of 7.4 Mev.

6.91 Mev level (2^+); The alpha-particle width is unknown; call it b_1 times the single particle limit. The radiation width is 0.03 electron volts $\pm$ 40 percent. [82].

7.12 Mev level (1^-): The alpha-particle width is unknown; call it b_2 times the single particle limit. The radiation width is 0.1 electron volts $\pm$ 60 percent [82].

9.58 Mev level (1^-); The alpha-particle width is 1.27 times the single particle limit, and the radiation width is 6×10^{-3} electron volts [83; 76].

13.09 Mev level (1^-); The alpha-particle width is 0.016 times the single particle limit and the radiation width is 28 electron volts in the thermonuclear energy region. Although this level is a long way away from the Gamow peak, it is a very special level in that it has large values of both the alpha-particle and electric dipole radiation widths. This seriously violates isotopic spin selection rules in O^{16}.

The $C^{12}(\alpha, \gamma)O^{16}$ reaction rate is therefore found to be

$$p_2 = (17b_1 + 360b_2 + 9) \times 10^{-14-139.3/T^{1/3}} \frac{n_1}{T^{2/3}} \text{ s}^{-1}, \tag{8.2.2}$$

$$= q_2 n_1, \tag{8.2.3}$$

$$\text{where } q_2 = \frac{(17b_1 + 360b_2 + 9)}{T^{2/3}} \times 10^{-14-139.3/T^{1/3}}. \tag{8.2.4}$$

8.2.1 Supplementary Notes: The $C^{12}(\alpha, \gamma)O^{16}$ reaction

Swann & Metzger [82] have given revised values for the radiation widths of the 6.91 and 7.12 Mev levels in O^{16}. With these slight changes the reaction rate given in the text is obtained.

J. K. Perring (unpublished) has developed the theory of the alpha-particle model of O^{16} to determine for which states in this model alpha-particle emission is allowed or forbidden. The 7.12 Mev level should have allowed emission and we may therefore expect that the alpha-particle reduced width is large; perhaps $b_2 \approx 1$. In this case, the capture in the tail of the 7.12 Mev level may be expected to dominate the reaction rate, although in fact the thermonuclear energy region is far enough from the resonance that a multi-level formula will eventually have to be used, taking into account interference between the 7.12 and 9.58 Mev levels.

Since the 7.12 Mev level is only slightly bound, it would be desirable to take into account the energy dependence of the Breit-Wigner denominator in the reaction rate for the $C^{12}(\alpha, \gamma)O^{16}$ reaction. When this is done we get to a good approximation:

$$p_2 = 1.1 \times 10^{-10-139.3/T^{1/3}} \frac{b_2 n_1}{T^{4/3}} \text{ s}^{-1}. \qquad (8.2.5)$$

8.3 The formation of neon

The oxygen nuclei which are formed can also capture an alpha-particle:

$$O^{16}(\alpha, \gamma)Ne^{20}. \qquad (8.3.1)$$

The excited states of the Ne^{20} nucleus lie at excitation energies of 1.63, 4.26, 4.97, and 5.81 Mev [84]. The Gamow peak would come in the vicinity of 5.0 Mev; hence the reaction is resonant provided the 4.97 Mev level can be formed by combining alpha-particles with O^{16}. This is possible only if the spin and parity of the 4.97 level are both odd or both even.[§]

The 1.63 Mev state is known to be 2^+; the spins and parities of the other excited

---

[§]Editor's note: These combinations of spin and parity are now known as *natural parity*.

states are unknown. However, light nuclei in this region have many properties which are successfully explained by the collective model developed for heavy nuclei [85]. Thus it is tempting to postulate that the first four excited states are precisely the four lowest lying stats which will be rotationally and vibrationally related to the ground state. One would expect 2^+ and 4^+ rotational states and the 2^+ and 3^+ vibrational states. The 3^+ state could not be formed by alpha-particles, but it is likely to lie at a higher energy than the other states. We shall assume that the 4.97 Mev state can be formed by alpha-particles, and that its spin and parity are likely to be 2^+ or 4^+.

The radiation width is bound to be ten or more orders of magnitude larger than the alpha-particle width of the 4.97 Mev state; hence only the latter width is needed for the resonant rate. Let the alpha-particle width of the 4.97 Mev state be c times the single particle limit, and let d be the product of the ratio of the Coulomb penetrability probability to that for s-wave alpha-particles times the statistical factor g in the Briet-Wigner formula (5.2.1). If the angular momentum of the incident alpha-particles is 0, 1, 2, 3, or 4, then $d = 1, 1.350, 0.465, 0.658,$ or 0.00459, respectively. The reaction rate is

$$p_3 = 6.34 \times 10^{-29-992/T} \frac{n_1 c d}{T^{3/2}} \text{ s}^{-1}, \qquad (8.3.2)$$

$$= q_3 n_1, \qquad (8.3.3)$$

$$\text{where } q_3 = 6.34 \times 10^{-29-992/T} \frac{c d}{T^{3/2}}. \qquad (8.3.4)$$

Still, further alpha-particles might be captured in reactions:

$$\text{Ne}^{20}(\alpha, \gamma)\text{Mg}^{24}, \qquad (8.3.5)$$

$$\text{Mg}^{24}(\alpha, \gamma)\text{Si}^{28}, \qquad (8.3.6)$$

and so on. However, these reactions are probably only of minor importance in stellar interiors, and they are not discussed in the following analysis.

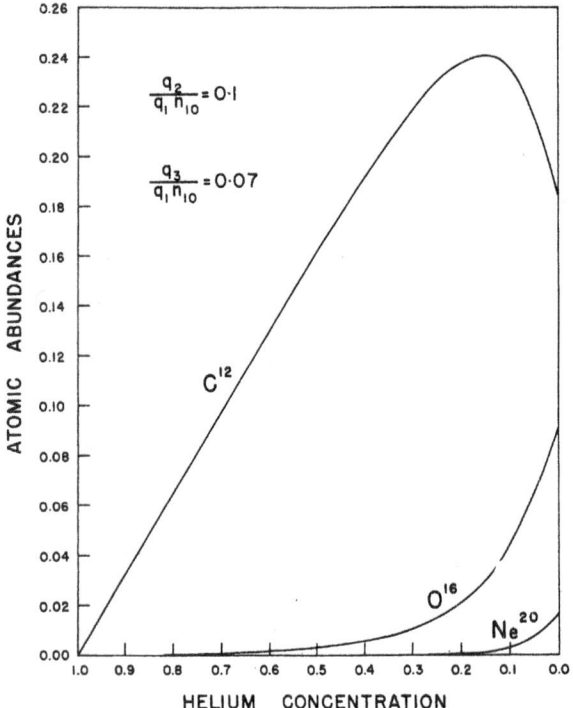

Figure 8.4.1: Solutions of the abundance equations for helium thermonuclear reactions with the parametric ratios $q_2/q_1 n_{10} = 0.1$; $q_3/q_1 n_{10} = 0.07$.

8.3.1 Supplementary Notes: The $O^{16}(\alpha, \gamma)Ne^{20}$ reaction

Gove, Litherland & Ferguson [86] have shown that the 4.97 Mev level in Ne^{20} is formed by a radiative transition from a 1^+ state of Ne^{20}, that it decays to ground less than five percent of the time, and that it does not have zero spin. This experiment limits the spin-parity assignments of the level to the possibilities $1^\pm$, $2^\pm$, or 3^+. Spin 1 is somewhat unlikely in view of the small number of transitions to the ground state. The presence of neon enhancements in some stars suggests then that the state is 2^+ and hence reaction 8.3.1 has a thermonuclear resonance.

84

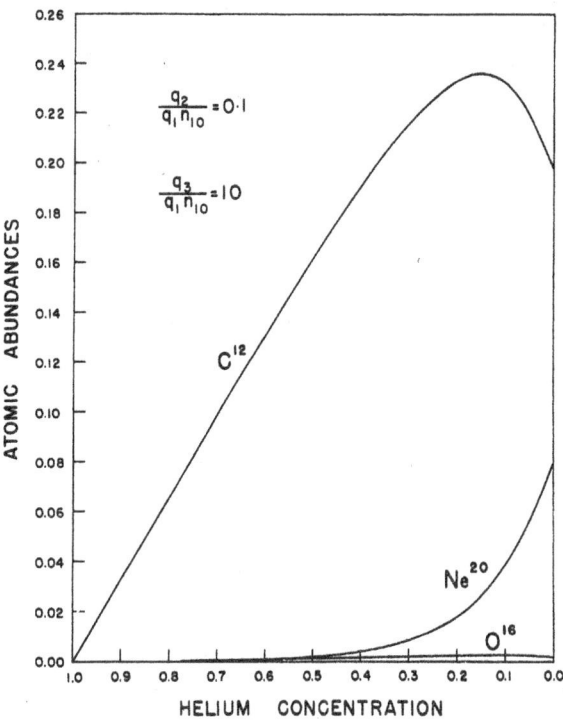

Figure 8.4.2: Solutions of the abundance equations for helium thermonuclear reactions with the parametric ratios $q_2/q_1 n_{10} = 0.1$; $q_3/q_1 n_{10} = 10$.

8.4 Products of the helium reactions

The hotter stars of the main sequence, with central temperatures of 17×10^6 °K and more, convert hydrogen to helium mainly by the carbon-nitrogen cycle. We have written the helium thermonuclear reaction rates in terms of functions $q(T)$ which depend only on nuclear constants and on the temperature. These functions appear in the differential equations which describe the changes in the abundances of He^4, C^{12}, O^{16}, and Ne^{20} as

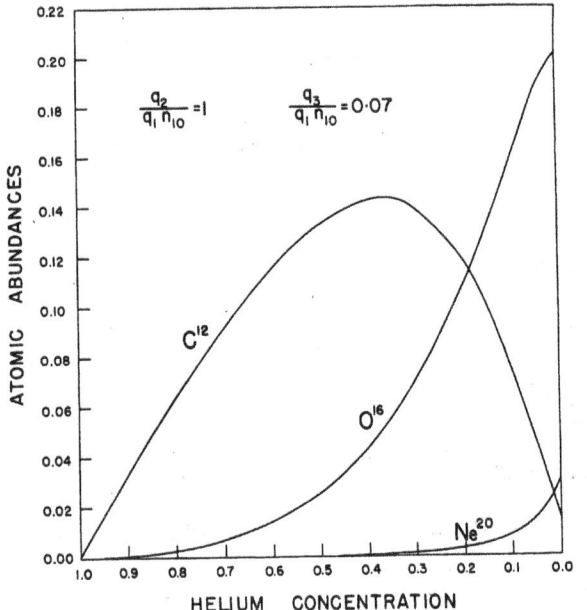

Figure 8.4.3: Solutions of the abundance equations for helium thermonuclear reactions with the parametric ratios $q_2/q_1 n_{10} = 1$; $q_3/q_1 n_{10} = 0.07$.

thermonuclear reactions take place in a helium gas. The equations are:

$$\frac{dn_1}{dt} = -3q_1 n_1^3 - q_2 n_1 n_3 - q_3 n_1 n_4,$$
$$\frac{dn_3}{dt} = q_1 n_1^3 - q_2 n_1 n_3,$$
$$\frac{dn_4}{dt} = q_2 n_1 n_3 - q_3 n_1 n_4,$$
$$\text{and} \quad \frac{dn_5}{dt} = q_3 n_1 n_4. \tag{8.4.1}$$

Here n_1, n_3, n_4, and n_5 are the number densities of He^4, C^{12}, O^{16}, and Ne^{20} nuclei, respectively. The destruction of Ne^{20} has been neglected. Numerical solutions to the above equations have been obtained by Hoyle [77], Nakagawa *et al.* [78], and Hayakawa, Hayashi, Imoto & Kikuchi [87]. All these solutions have been obtained under the simple assumption that the temperature and density of the gas remain constant. The writer gives here some independent solutions of his own, presented in a way which appears to bring out more clearly the dependence of these solutions on unknown nuclear constants

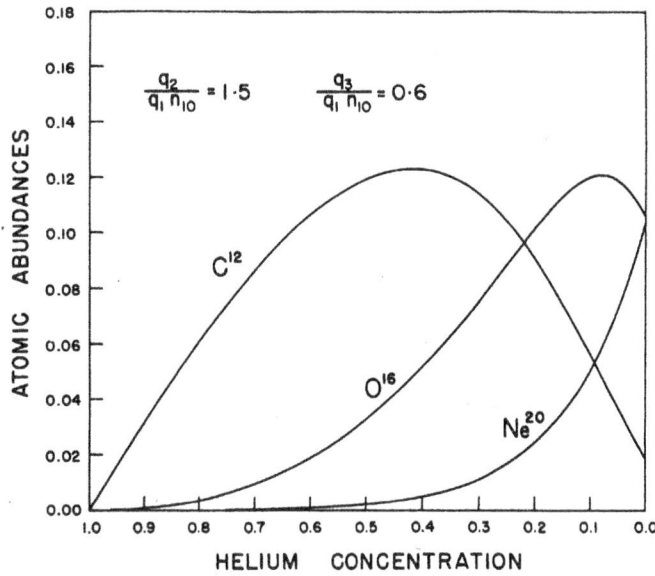

Figure 8.4.4: Solutions of the abundance equations for helium thermonuclear reactions with the parametric ratios $q_2/q_1 n_{10} = 1.5$; $q_3/q_1 n_{10} = 0.6$.

and on the conditions in the helium gas. The solutions depend on two parametric ratios:

$$\frac{q_2}{q_1 n_{10}} = (2.3b_1 + 52b_2 + 1.4) \times 10^{-11 - 139.3/T^{1/3} + 1886/T} \frac{T^{7/3}}{\rho}, \qquad (8.4.2)$$

$$\text{and} \quad \frac{q_3}{q_1 n_{10}} = 4.4 \times 10^{-5 + 892/T} \frac{cT^{3/2}}{\rho}. \qquad (8.4.3)$$

Here n_{10} is the initial number density of He^4 nuclei, and it has been assumed that the 4.97 Mev level of Ne^{20} is 2^+. If the latter level is 4^+, the ratio $q_3/q_1 n_{10}$ must be decreased by a factor 100. It should be noticed that both parametric ratios are decreasing functions of temperature. Therefore, if one were to adopt the condition that the helium reactions take place with a constant rate of energy generation, the temperature would be an increasing function of time and the parametric ratios a decreasing function of time. The assumption of constant temperature and density means the ratios remain constant.

Solutions corresponding to selected parametric ratios are shown in Figures 8.4.1 through 8.4.9. It may be seen that for $q_2/q_1 n_{10} = 0.1$, C^{12} is the main product of

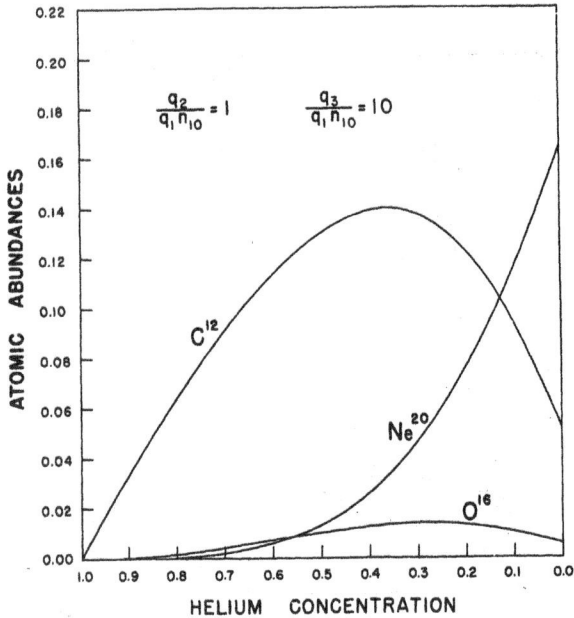

Figure 8.4.5: Solutions of the abundance equations for helium thermonuclear reactions with the parametric ratios $q_2/q_1 n_{10} = 1$; $q_3/q_1 n_{10} = 10$.

the reactions and remains in large abundance even when the helium is exhausted. For larger values of this ratio up to $q_2/q_1 n_{10} = 10$, the C^{12} abundance is large to begin with but becomes small when the helium is exhausted. Even when $q_2/q_1 n_{10} = 100$, C^{12} is the principal product during the consumption of the first five percent of the helium. This general feature of the reactions is of importance in the later discussion.

It may also be seen in these figures that for $q_3/q_1 n_{10} = 0.07$ the main endproduct of the reactions (except possibly for C^{12}) is O^{16}. However, for the higher values shown for the ratio, the main end-product is Ne^{20}. An intermediate case is shown in Figure 8.4.4.

Hence the question of Ne^{20} formation depends critically on the alpha-particle width of the 4.97 Mev level in the Ne^{20}. Let us consider what values of these parameters are likely to arise under different conditions. Perhaps the most useful parameter to employ is the rate of energy generation in the gas at the start of the helium reactions, when essentially the only product is C^{12}. This energy generation rate can be connected fairly readily to the luminosities of the different stars even if one does not know the particular

88

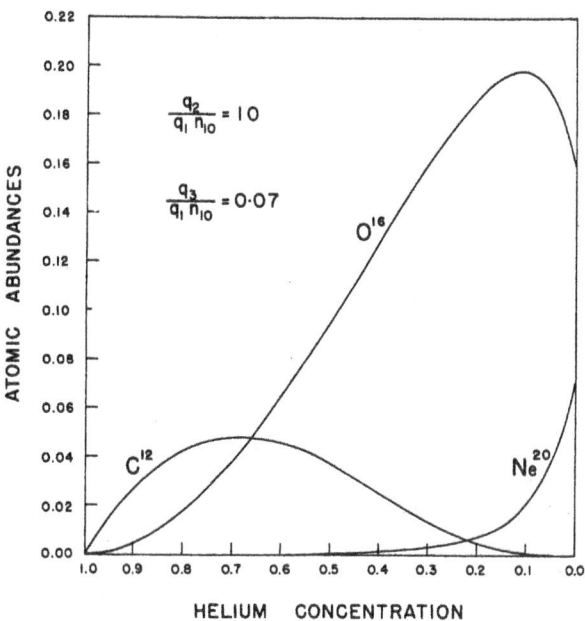

Figure 8.4.6: Solutions of the abundance equations for helium thermonuclear reactions with the parametric ratios $q_2/q_1 n_{10} = 10$; $q_3/q_1 n_{10} = 0.07$.

combinations of temperature and density responsible for it. The initial rate of energy generation is

$$\epsilon = 4 \times 10^{64} q_1 \rho^2 \text{ ergs/g s.} \tag{8.4.4}$$

The density corresponding to various combinations of ϵ and T are shown in the following table:

ϵ \ T	100	120	140
10^2	6.4×10^4	2.2×10^3	2.1×10^2
10^4	6.4×10^5	2.2×10^4	2.1×10^3
10^6	6.4×10^6	2.2×10^5	2.2×10^4

Here T is in the usual units of 10^6 °K and ρ is in g/cm^3. The globular cluster stars of Population II can be expected to consume helium at $\epsilon \sim 10^4$ ergs/g s.

The following two tables show the values of the parametric ratios as functions of ϵ

89

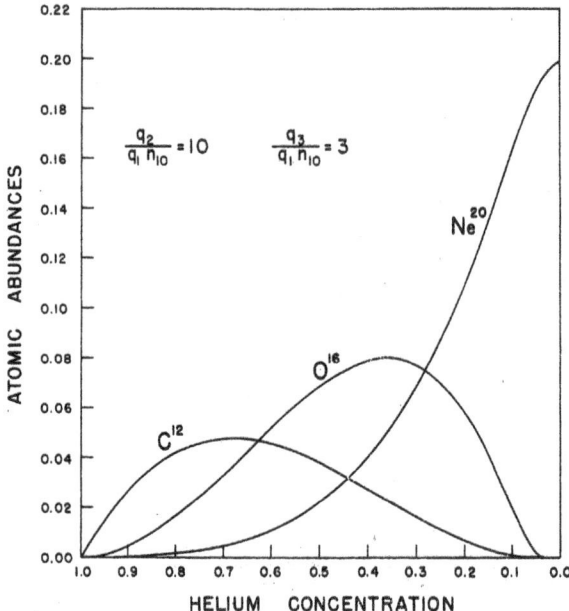

Figure 8.4.7: Solutions of the abundance equations for helium thermonuclear reactions with the parametric ratios $q_2/q_1 n_{10} = 10$; $q_3/q_1 n_{10} = 3$.

and T. First, the ratio $q_2/q_1 n_{10}$:

T ϵ	100	120	140
10^2	$1.1b_1 + 22b_2 + 0.6$	$1.9b_1 + 42b_2 + 1.1$	$4.8b_1 + 110b_2 + 2.8$
10^4	$0.11b_1 + 2.2b_2 + .06$	$0.19b_1 + 4.2b_2 + .11$	$0.48b_1 + 11b_2 + .28$
10^6	$.011b_1 + .22b_2 + .006$	$.019b_1 + .42b_2 + .011$	$.048b_1 + 1.1b_2 + .028$

The ratio $q_3/q_1 n_{10}$:

T ϵ	100	120	140
10^2	$600\ c$	$700\ c$	$800\ c$
10^4	$60\ c$	$70\ c$	$80\ c$
10^6	$6\ c$	$7\ c$	$8\ c$

It may be seen from these tables that for high rates of energy generation, the main product will be C^{12}. O^{16} formation would be favored for cases in which $b_1 = b_2 = 1$.

90

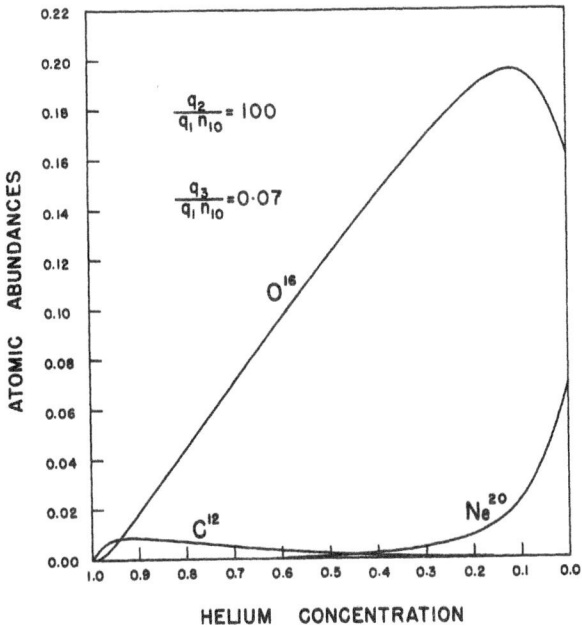

Figure 8.4.8: Solutions of the abundance equations for helium thermonuclear reactions with the parametric ratios $q_2/q_1 n_{10} = 100$; $q_3/q_1 n_{10} = 0.07$.

However, on a statistical basis, it is much more likely that these reduced widths are much smaller, say $b_1 \simeq b_2 \simeq 0.01$. In this case most of the value of $q_2/q_1 n_{10}$ comes from the third term which is known. In any case, values of $q_2/q_1 n_{10}$ as high as 100 can be obtained only if the 7.12 Mev level in O^{16} has a reduced width comparable to the single particle limit and if the rate of energy generation is very low. Hence it appears to be a safe conclusion that the main product of the helium thermonuclear reactions during the expansion of the core in globular cluster stars will be C^{12}. If the 4.97 Mev level of Ne^{20} is 2^+ as assumed previously, then it appears that stars can produce much Ne^{20} during the last stages of helium consumption for quite small values of the reduced alpha-particle width, say $c \simeq 0.01$. However, if the level is 4^+, then substantial amounts of Ne^{20} are produced only if the reduced width is quite large, $c \sim 1$. However, if the 4.97 Mev level cannot be formed by combining alpha-particles with O^{16}, then neon production by helium thermonuclear reactions is negligible because the nonresonant reaction rate

91

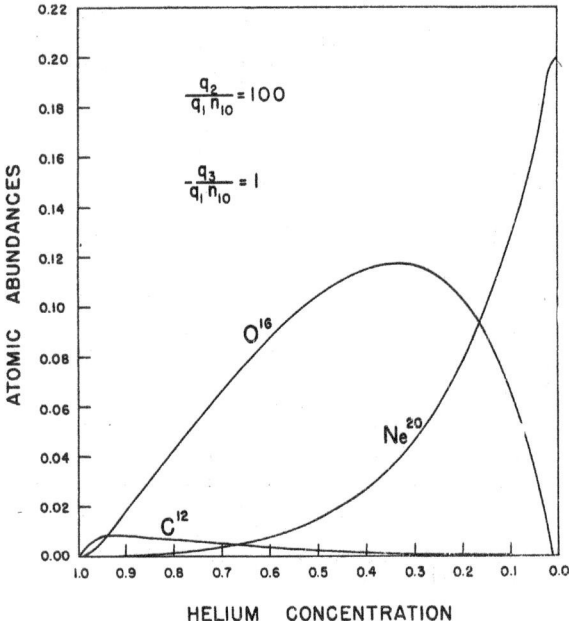

Figure 8.4.9: Solutions of the abundance equations for helium thermonuclear reactions with the parametric ratios $q_2/q_1 n_{10} = 100$; $q_3/q_1 n_{10} = 1$.

would be very much less than the resonant rate given here.

8.5 Carbon stars

We will now consider certain classes of stars which have abundance anomalies, possibly connected with the helium thermonuclear reactions. We have already mentioned the carbon stars — those of spectral classes R and N — which have an excess of carbon over oxygen in their atmospheres. These stars are strongly concentrated to the plane of the galaxy. Therefore they are members of Population I, and they may include members with very high rates of energy generation.

We have also seen that one possible explanation for some of these stars is that their atmospheric materials have been exposed to high temperatures long enough to destroy most of the oxygen. However, this explanation does not explain all the stars (or any of them very well), since many of the stars have ratios of C^{12} to C^{13} abundances much larger than the carbon-cycle value of four.

92

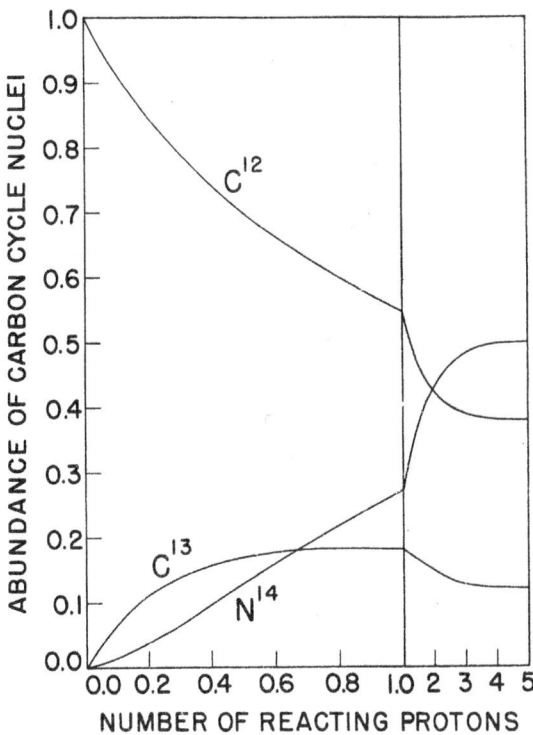

Figure 8.5.1: Abundances of carbon-cycle constituents as a function of the number of protons interacting with a C^{12} gas. N^{15} abundances are too small to be shown. The final equilibrium abundance of N^{14} should be much larger than shown, since this figure was calculated for old carbon-cycle reaction rates.

It appears to be more satisfactory to suggest that most if not all of the carbon stars have mixed into the surface, additional carbon formed by helium thermonuclear reactions in the core. The mechanism of mixing is not clear; extensive convection zones may well have been formed several times for short periods if the cores of those stars have repeatedly reached the point of instability. However, it is not clear whether these convection zones will extend to the surface. Some red giant stars eject large quantities of matter into space, and it is possible that this process uncovers the material which had been convected part way to the surface. There are many puzzles to be cleared up in this question of mixing to the surface.

If the carbon stars have been enriched in this element by mixing the products of

93

helium thermonuclear reactions into the surface layers, then it is quite possible that some of them will have different isotopic abundance ratios of C^{12} to C^{13}. Those with values other than four will be cases in which the carbon has spent very little time in high temperature regions. However, it should not necessarily be assumed that those with ratios of four have achieved carbon-cycle equilibrium proportions in which N^{14} is by far the most abundant constituent. The situation is illustrated in Figure 8.5.1, which shows how rapidly the various carbon-cycle nuclei approach equilibrium abundances as protons are allowed to interact with an initially pure C^{12} gas. It maybe seen that the C^{13} abundance rapidly approaches its equilibrium abundance relative to C^{12} and is fairly constant thereafter, while the N^{14} equilibrium abundance builds up slowly. If the diagram had been properly computed using the reaction rate constants given in Table 6.1, the final N^{14} equilibrium abundance would be much higher than shown, but the early parts of the curves would hardly be affected.

Thus we see that the observation that the C^{12} to C^{13} abundance ratio is four is no guarantee that carbon-cycle equilibrium proportions have been attained. Indeed, it may well be that, in many cases, the attainment of equilibrium would reduce the C^{12} and C^{13} abundances so much that they would no longer be in excess over that of O^{16}, and the star would still have an H or S spectrum.

Stars of spectral class S also show unusually large amounts of carbon, but in this case the oxygen abundance has not been exceeded, and the spectrum still contains oxide bands. Other kinds of stars showing large abundances of carbon are the CH stars, the Ba II stars, and the R Coronæ Borealis variable stars, as well as those to be discussed in the following sections.

8.6   Wolf-Rayet stars

The spectra of Wolf-Rayet stars exhibit prominent broad emission lines which are usually interpreted as arising from gaseous layers having large velocities of ejection from the stars. The W stars are usually deficient in hydrogen and have abnormally large abundances of carbon or nitrogen. Aller [88] states that in a typical WC star the ratio of helium, carbon, and oxygen abundances is 50:3:1. The WN stars appear to be similar, but with the carbon mainly replaced by nitrogen with the nitrogen to carbon ratio having the typical carbon-cycle ratio of 10 or 20. These numbers result from an analysis in which the Wolf-Rayet atmosphere is assumed to be in thermal equilibrium, and they must be treated with great caution. Zanstra & Weenan [89] have shown that if one considers the WC stars to have an extended atmosphere excited by diluted radiation, then the helium to carbon ratio is about three in three examples studied. Weenan [90] argues that even with the opposite extreme assumption of thermal equilibrium, the helium to carbon ratio is considerably lower than Aller's values in the three stars studied.

The W stars therefore appear to be horizontal branch objects in an advanced stage of evolution in which much of the envelope has been cast off into space, and in which the surface layer contains a considerable admixture of the products of helium thermonuclear reactions. The carbon and nitrogen sequences seem to represent extremes in which the C^{12} produced in the core has not or has reacted with envelope hydrogen, respectively. Some intermediate cases exist in which the carbon and nitrogen are of comparable abundance, and in which either the carbon-cycle reactions did not go far enough for equilibrium to be reached, or only did so for part of the material from the core which is now in the surface.

8.7 Helium stars

These are hot, horizontal branch stars whose composition appears to be similar to that of the Wolf-Rayet stars. Greenstein [91] mentions cases in which such stars have strong helium and carbon lines, but no hydrogen lines. There are also cases in which helium, nitrogen, and neon give strong lines and hydrogen gives weak lines. Thackeray [92] has observed a helium star in which no lines of hydrogen or oxygen are visible, but in which helium, carbon, and neon give strong lines. He remarks particularly about the unusual strength of the neon lines and concludes that the oxygen abundance must be less than that of carbon. It also seems likely that oxygen is of smaller abundance than carbon in the other cases mentioned by Greenstein.

The presence of neon in these stars appears to be astrophysical evidence that the 4.97 Mev level in Ne^{20} can be formed in the $O^{16}(\alpha, \gamma)Ne^{20}$ reaction and that it has a spin-dependent value of the reduced alpha-particle width of the order of magnitude indicated in Section 8.3.

8.8   Helium reactions following mixing

It has been suggested in Section 7.3 that hydrogen from the envelope is mixed into the core when the core expands as helium reactions start in a globular cluster star at the tip of the giant branch. Following the mixing there may be a small but significant content of carbon-cycle products in the core which can take part in later helium thermonuclear reactions. The reactions of interest are:

$$C^{13}(\alpha, n)O^{16}, \tag{8.8.1}$$

$$N^{14}(\alpha, \gamma)F^{18}, \tag{8.8.2}$$

$$\text{followed by} \quad F^{18}(\beta^+\nu)O^{18}, \tag{8.8.3}$$

$$\text{and} \quad N^{15}(\alpha, \gamma)F^{19}. \tag{8.8.4}$$

The reaction (8.8.1) will be of particular interest to us in the following chapters. The neutrons produced are slowed down to thermal equilibrium with their surroundings (having an energy of about 10 kev at temperatures near 100×10^6 °K). The neutrons cannot be captured by the helium since He^5 is unstable, but they will be captured by other nuclei present in proportion to their abundances and capture cross sections near 10 kev. We shall consider these quantities in the following chapters.

Fowler, Burbidge, & Burbidge [93] have suggested that the reaction

$$Ne^{21}(\alpha, n)Mg^{24} \tag{8.8.5}$$

gives an important source of neutrons in the stellar interior, especially if the Ne^{20} can have been largely converted to Ne^{21} during hydrogen consumption in the shell source at high temperatures. We saw in Section 6.6 that it is questionable whether this would occur.

96

Another source of neutrons:

$$O^{17}(\alpha, n)Ne^{20} \tag{8.8.6}$$

is also likely to be unimportant because the O^{17} is rapidly destroyed in hydrogen thermonuclear reactions.

The reaction rate of the $C^{13}(\alpha, n)O^{16}$ reaction has been estimated in two ways which give upper and lower limits.

The cross section for the inverse reaction $O^{16}(n, \alpha)C^{13}$ has been measured by Seitz and Huber [94]. The excitation energy in O^{17} involved in their experiment lies above the thermonuclear energy region. However, they have observed resonances which do not show up in neutron scattering (apparently due to lack of energy resolution). It is possible to make an approximate estimate of the contribution of these levels to the $O^{16}(n, \alpha)C^{13}$ reaction at lower energies using the Breit-Wigner single-level formula. The cross section for the $C^{13}(\alpha, n)O^{16}$ reaction then can be determined using the nuclear reciprocity theorems. This gives a lower limit to the reaction cross section in the thermonuclear energy region.

An effective upper limit to the cross section in the thermonuclear energy region is obtained by assuming that the 6.37 Mev level in O^{17} has a reduced alpha-particle width equal to the single-particle limit. This level lies closest to the thermonuclear energy region; it has a reasonably large neutron width (120 kev); and it can be formed by p-wave alpha-particles bombarding C^{13}.

The reaction rate so obtained is

$$p = (3 + 130f) \times 10^{16-140/T^{1/3}} \frac{\rho x_1}{T^{2/3}} \text{ s}^{-1}, \tag{8.8.7}$$

where f is the reduced alpha-particle width of the 6.37 Mev level in O^{17} expressed as a fraction of the single particle limit.

This relation shows that C^{13} will start reacting with helium at a temperature of about 80×10^6 °K and will go at a fairly respectable rate at 100×10^6 °K. Neutron capture effects will be calculated for a temperature of 100×10^6 °K in the subsequent discussion. If the $Ne^{21}(\alpha, n)$ reaction is important, it will take place at a temperature of about 160×10^6 °K [93].

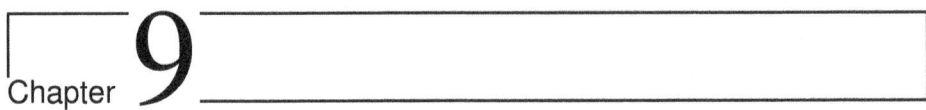

Chapter 9

The Abundances of the Elements

It is a very difficult problem to determine the relative abundances of the elements in the universe. We have mentioned the systematic differences which exist between Populations I and II, and we have seen that certain stars have very peculiar abundances indeed. It is very difficult to make accurate determinations of abundances in stars, even in the sun. It is not much easier to analyze other objects in the solar system, including the earth. A comprehensive discussion of these problems is given by Aller in a forthcoming book*. Some brief comments on these problems are given here.

9.1 Geochemical abundances

The volatile elements are very greatly depleted in the earth. Hydrogen and helium have been lost, together with most of the rare gases. Carbon and nitrogen were lost as methane and ammonia, and much of the oxygen as water vapor. The remaining elements have been concentrated in different phases in the earth. For example, the most abundant elements in the earth's crust are oxygen, silicon, aluminum, hydrogen, sodium, calcium, iron, magnesium, and potassium, in that order. But for the earth, as a whole, the most abundant elements are oxygen, magnesium, iron, silicon, sulfur, nickel, aluminum, calcium, and sodium, in that order. The rarer elements often concentrate even more markedly in different phases. Thus, although the earth gives us a good idea of what the

---
*Editor's Note: Published as [95].

more abundant nonvolatile elements are, it is extremely difficult to make good estimates of the terrestrial average abundances of most elements. On the other hand, we can obtain very accurate values of the relative isotopic composition of the elements from analyses of terrestrial materials.

9.2 Meteorite abundances

Most meteors entering the earth's atmosphere appear to be debris from comets; they are quickly broken up and consumed in the upper atmosphere. However, some of the brighter meteors survive their passage through the atmosphere and can be recovered as meteorites on the ground. Most of these appear to be fragments of former bodies in the asteroid belt which have been broken up by collisions.

There are a great many different forms of the meteorites, which can roughly be described as irons, stony-irons, and stones. The most interesting, from the point of view of element abundances, are the chrondrite meteorites. These are very numerous and consist of conglomerates of broken and partly melted minerals containing inclusions of round chrondules and smaller irregular pieces of material. The chrondules appear to have been molten droplets which have solidified in free space. Urey [96] concludes that the chrondrites are objects which result from a melting process, a cooling, a shattering process, a reaccumulation into a larger body, followed by another shattering process. Urey [97] proposed that the chrondrites themselves give a good sample of the nonvolatile constituents of the solar system. Although there are individual differences in composition between different chrondrites, Urey's view has led to many, very useful results. Certainly the chrondrites are far more homogeneous bodies than one can ever find on the earth.

9.3 Stellar abundances

Despite the great difficulties associated with the determination of the chemical compositions of the stars, astrophysicists have developed refined theories for interpreting stellar spectra, and quite good abundances are now available for the more abundant elements. The situation has recently been reviewed by Greenstein [98]. The better-determined abundances of nonvolatile elements agree quite well with those determined

99

in meteorites. The abundances of volatile elements like H, He, C, N, O, and Ne must be determined from stellar sources; the situation regarding most of these is fair, but complicated by abundance differences in several sources.

In this connection it may be mentioned that there is some preliminary and still inconclusive evidence that the ratios of nitrogen to carbon and of neon to oxygen are increasing functions of the time of formation of objects in our galaxy (L.H. Aller, private communication). This is the sort of behavior one would expect if, in the less massive stars now passing through advanced evolutionary stages, the carbon produced in the core has a greater chance of then being processed in the carbon cycle, and if the core evolves at somewhat lower temperatures better suited to the production of neon.

Goldberg, Müller, & Aller [60] have recently made improved determinations of the abundances of many elements in the sun. Among the more significant of their findings are these: The relative abundances of C, N, O, and H are 3.6, 0.95, 10, and 10,000, respectively. Iron is slightly less abundant and chromium more abundant than obtained in meteorite analyses. Lead is much more abundant in the sun than in the meteorites.

9.4 The semi-empirical abundances of Suess and Urey

For some considerable time it has been known that certain properties of nuclear matter were reflected in the abundances of the nuclides. For example, the large abundances of nuclides with certain "magic numbers" of neutrons and protons were used as early arguments for the existence of closed nucleon shells in nuclei. The argument has recently been reversed: Suess [99; 100; 101] and Suess & Urey [102] argue that the most regular effect of nuclear properties on nuclide abundances is the steady progression of the abundances of the nuclides with odd mass numbers. Therefore they have made preliminary assignments of abundances of the elements from terrestrial, meteorite, and stellar analyses. They have then adjusted these so that the abundances of nuclides with odd mass numbers fall on a smooth curve. They have been greatly helped in this process by the fact that many elements contain two isotopes with odd mass numbers; these establish the slope of the abundance curve at many places.

The writer has modified the Suess-Urey abundances somewhat to take account of the

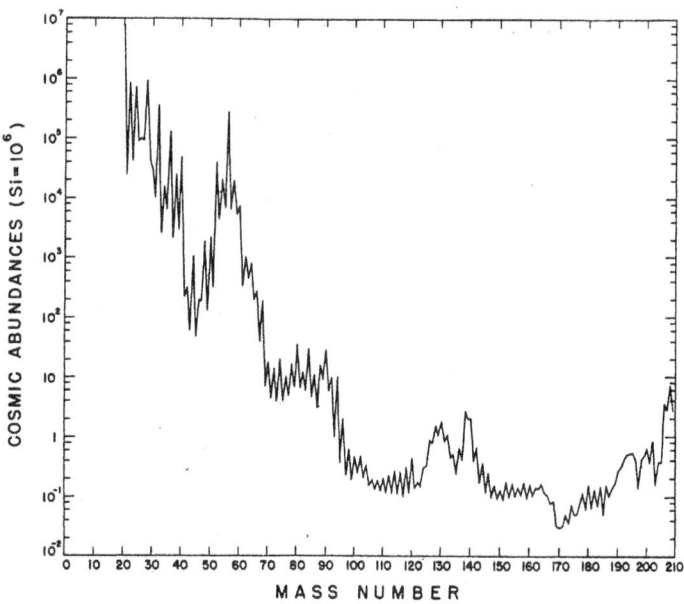

Figure 9.4.1: The cosmic abundances of Suess & Urey [102] plotted as functions of mass number. This curve has been modified in the iron peak and lead regions to take partial account of the new solar abundance determinations of Goldberg, Müller & Aller [60].

new solar abundances of Goldberg, Müller & Aller [60], particularly in the region of the iron abundance peak and in the region of lead. The resulting curve as a function of mass number is shown in Figure 9.4.1.

The Suess-Urey semi-empirical abundances have been most useful in interpreting the abundances of the elements as resulting from the superposition of several mechanisms of nuclear synthesis. We shall perform such analyses later.

Chapter 10

Neutron-Capture Cross Sections

We shall see that most neutrons injected into a stellar interior will be captured by heavy elements or by N^{14}. Each $C^{13}(\alpha, n)O^{16}$ reaction releases 2.2 Mev of energy. If the neutron is captured by a heavy nucleus, then about another 8 Mev of energy is released. If the capture is by N^{14}, then a further 3.6 Mev is released. In a typical globular cluster star, in an advanced stage of evolution, the energy generation in the core is of the order of 10^4 ergs/g s. If 20 neutrons are injected per initial silicon atom into some of this matter at a rate sufficient to maintain this energy generation, then the period of the injection would cover 100 years. We shall see that a typical heavy nucleus may capture about 50 neutrons in this period, giving a mean time between neutron captures of about 2 years. However, this is likely to be too short a time because it is doubtful that the neutron production will account for all the energy generation in the star for such a short period, and it is also likely that the time required to capture 20 neutrons per initial silicon atom will be extended owing to capture in N^{14}. Therefore, as an order of magnitude we may assume that there will be 10 to 100 years between neutron captures in a typical heavy nucleus. This process will be referred to as neutron capture on a slow time scale.

Many of the products of the neutron capture are unstable to beta emission. If these products have beta-decay half-lives longer than the half-lives for destruction by neutron capture, they can be regarded as stable in the neutron-capture process, even though, in a strict sense a small proportion of such nuclei will decay before capturing a neutron. If the half-life for beta decay is shorter than the half-life for neutron destruction, then this decay can be assumed to take place before capture of a neutron, even though a small fraction of such nuclei may not have decayed. In this way, it is possible to determine a nearly unique neutron-capture path which will be followed by nuclei as they capture successive neutrons. If a nucleus is not initially on this capture path, it will approach the path when it starts capturing neutrons and then will follow it. In this work it has been customary to assume that the neutron-destruction half-life and the beta half-life are equal for a beta-decay half-life of 10 years and a neutron-capture cross section (at 11 kev) of 1 barn. The neutron-destruction half-life varies inversely proportionally to the capture cross section.

We will now consider briefly the methods which the writer uses to compute capture cross sections in the kilovolt region for nuclides on the main neutron-capture path. Only a very few capture cross sections have been measured in this energy region, so it is necessary to resort to a theory in order to obtain most values.

10.1   Nuclides with large level spacings

Some nuclides (mainly with low mass numbers) do not have any neutron scattering resonances at energies below 100 kev. In such cases, the cross section in the kilovolt region can be obtained by extrapolation from measurements at laboratory thermal energies. The cross section obeys the $1/v$ law to a good approximation. Now at a temperature of 100×10^6 °K, a $1/v$ cross section averaged over the Maxwell distribution of velocities is equal to the actual cross section at 11 kev. Therefore, these cross sections have been computed at 11 kev. It turns out that in the case where there are many levels, the average cross section also varies approximately as $1/v$, so, in that case as well, the cross sections have been calculated at 11 kev.

In some nuclides there are only one or two neutron-scattering resonances below 100 kev, but the lowest lying level is still at 30 kev or higher. In such cases, the procedure has been to assume that the thermal-capture cross section is entirely due to the lowest lying level, provided the scattering shows the level to be formed by s-wave neutron capture. This evaluates the product of the neutron and radiation widths which appears in the Breit-Wigner formula. The cross section at 11 kev can then be computed using this

103

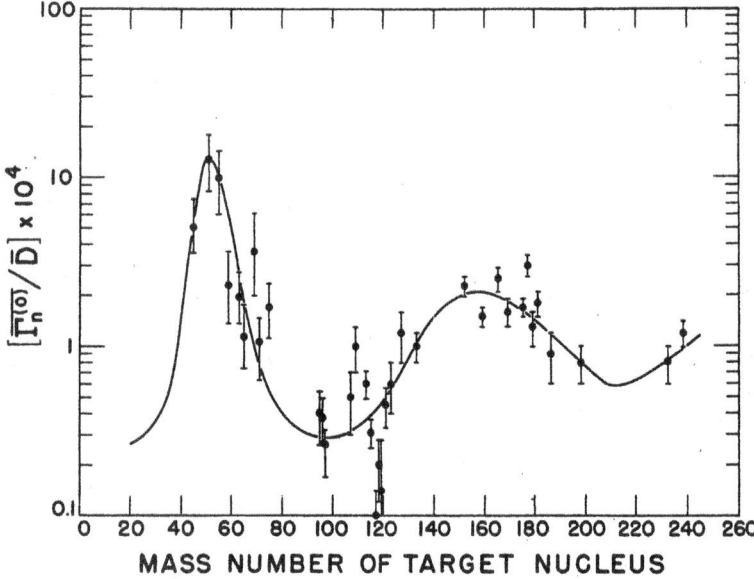

Figure 10.2.1: The s-wave neutron strength function, plotted as the average neutron width at 1 electron volt divided by the average level spacing for levels of a given spin and parity.

formula, assuming the total width of the resonance to be small compared to its distance from 11 kev.

Still other nuclides have one or two scattering resonances between zero and 30 kev. In these cases it is necessary to estimate the radiation and neutron widths of these levels and to integrate the resulting capture cross sections over the Maxwell spectrum to get an average value. The neutron width is usually large enough to be measured and the radiation width can be estimated if necessary from a semi-empirical formulation of the writer [103]. Such radiation widths have probable errors of about 40 percent.

10.2 Nuclides with small level spacings

We now pass over to the case where there are many resonance levels in the region of the peak of the Maxwell distribution of energies. In this case it is necessary to estimate the capture cross section averaged over the resonances. The formula for such an average

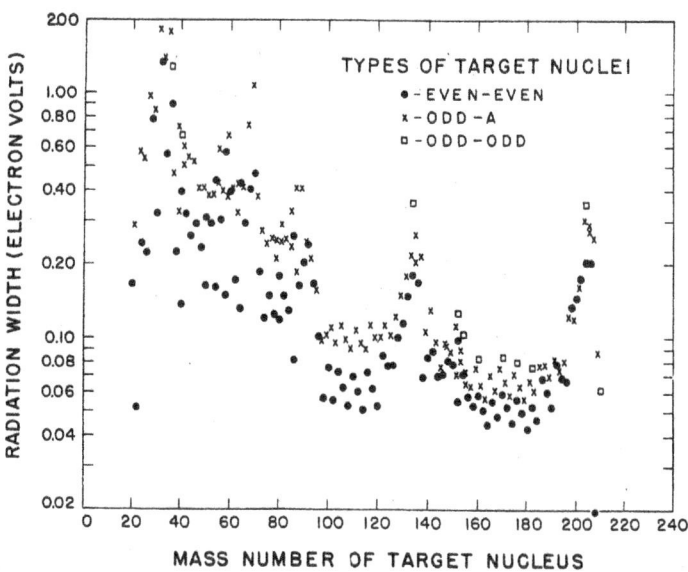

Figure 10.2.2: Radiation widths in compound nuclei formed when neutrons are captured by nuclei on the main capture path or close to it.

cross section is:

$$\bar{\sigma}(n,\gamma) = 2\pi^2\lambda^2 \sum_{J,\pi} \frac{\epsilon_{j\ell}^J(2J+1)}{(2(2I+1))} \frac{\bar{\Gamma}_n\bar{\Gamma}_\gamma}{(\bar{\Gamma}_n+\bar{\Gamma}_\gamma)\bar{D}} V\left[\frac{\bar{\Gamma}_\gamma}{\bar{\Gamma}_n}\right]. \qquad (10.2.1)$$

Here the summation extends over all levels of spin J and parity π in the vicinity of 11 kev. The statistical factor is defined as:

$$\epsilon_{j\ell}^J = 2 \text{ for } |J-\ell| \leq I \pm \frac{1}{2} \leq J+\ell,$$

$$= 1 \text{ for } |J-\ell| \leq \text{ only one of } I \pm \frac{1}{2} \leq J+\ell,$$

$$= 0 \text{ otherwise,}$$

where ℓ is the angular momentum of the incoming neutron.

The average neutron width, $\bar{\Gamma}_n$, must be calculated from a relation like (5.3.1). In the vicinity of 11 kev, it is necessary to take into account capture by both s-wave and p-wave neutrons. Equation (5.3.1) then demands knowledge of the s-wave and p-wave neutron strength functions $\gamma_n^2/\bar{D}$, well as the average level spacing $\bar{D}$. In the cloudy crystal ball

105

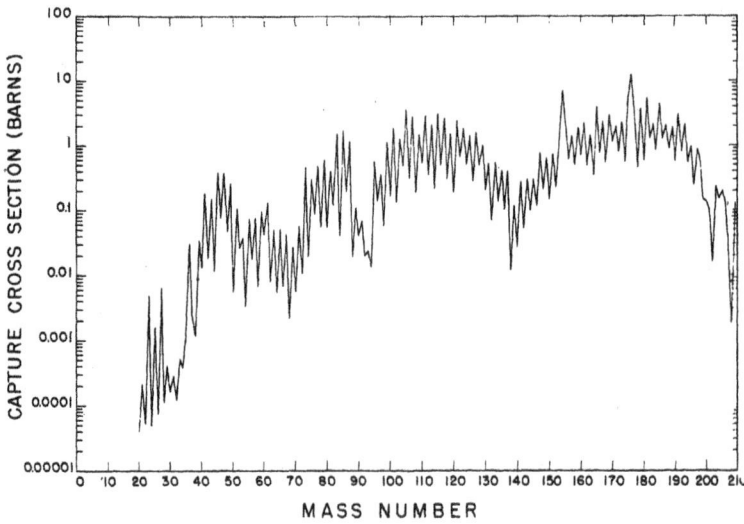

Figure 10.2.3: Neutron-capture cross sections at 11 kev computed by the methods outlined in Chapter 10. There is a numerical error in these computations; for heavier nuclei the cross sections should be increased by roughly a factor of 2.

nuclear model of Feshbach, Porter & Weisskopf [104], the strength function is peaked at certain mass numbers and has deep valleys at intermediate mass numbers. A plot of the s-wave strength function is shown in Figure 10.2.1, in which the data are taken from a compilation by Weisskopf [105], and the line is an empirical fit to the data which is consistent with the theory of Feshbach et al. [104]. The p-wave strength function is not known experimentally, and it has been necessary to assume a curve similar to Figure 10.2.1, but with peaks at intermediate mass numbers as predicted by the cloudy crystal ball model and with mass number widths interpolated and extrapolated from the curve in Figure 10.2.1.

The level spacings $\bar{D}$ were calculated from the theory of T. D. Newton [66] with improved values of the constants (quoted by Cameron [103]). This level-density formula gives values which have an average error of a factor three, thus representing a very significant advance over previous level-density formulas. There is some hope that further significant improvements can be made in this formula, and an attempt will be made to do this in the near future.

106

The level-density formula depends on the excitation energy in the compound nucleus, which is simply the binding energy of the neutron plus 11 kev. In the past the writer has used measured values of atomic masses, where these appeared to be good, in calculating the neutron binding energy. Where mass values are poor or nonexistent, the writer has taken neutron binding energies from the empirical formula of Levy [106]*, as tabulated by Riddell [108]. However, the writer has recently given the semi-empirical atomic mass formula a major overhaul and has devised empirical shell and pairing corrections to it; the resulting formula reproduces measured masses with median errors of about 300 kev [109]. These values will be used in future computations.

Average radiation widths, $\bar{\Gamma}_\gamma$, have been computed for nuclei of interest from the semi-empirical formula mentioned in the previous section [103]. The values obtained for nuclei which are possible members of the neutron-capture path are shown in Figure 10.2.2. It may be seen that there is a general downward trend with mass number, on which is superposed peaks immediately below closed neutron shells. The only remaining quantity in equation (10.2.1) to be discussed is the quantity $V\frac{\bar{\Gamma}_\gamma}{\bar{\Gamma}_n}$. This corrects for the fact that we have made independent use of the average neutron and radiation widths in the rest of the formula, whereas, in fact, we should have averaged $\frac{\Gamma_n\Gamma_\gamma}{(\Gamma_n+\Gamma_\gamma)}$ directly. The radiation widths are constant to within a few percent for different levels, but as we have seen from equation (5.3.8), the neutron widths can have an enormous variation, there being many small widths and a few large ones. If we assume a width variation as given by equation (5.3.8), then the correction factor is:

$$V(y) = (1+y)\left[1 - 1.772\left(\frac{y}{2}\right)^{1/2} e^{y/2}\left\{1 - H(z)\left[\left(\frac{y}{2}\right)^{1/2}\right]\right\}\right], \qquad (10.2.2)$$

$$\text{where } H(z) = \frac{2}{\pi^{1/2}}\int_0^z e^{-a^2}da. \qquad (10.2.3)$$

This correction factor turns out not to be very important; it lies between unity for $y = 0$ or $y = \infty$ and about 0.66 for $y \sim 1$.

Neutron-capture cross sections computed by these methods are shown in Figure 10.2.3. Unfortunately, a numerical error was made in these calculations which was not

---

*Editor's Note: Later published in a peer-reviewed journal [107].

discovered until further computations based on these cross sections had been made. The capture cross sections of heavier nuclei should be about a factor 2 larger than shown. This error will have little effect on the nature of the abundance changes which result from neutron capture, but it is important when one considers the competition of the $N^{14}(n,p)C^{14}$ reaction for the neutrons.

There are many measurements of neutron-absorption cross sections which have been made in fast reactors. In these measurements, substances have been exposed to a spectrum of neutrons which is very different from the Maxwell distribution in which we are interested. Most measurements have been done with neutron spectra peaking in the 0.1 to 1 Mev region; some have been done with spectra giving large contributions in the 11 kev region but peaking below it. In order to make a proper evaluation of these measurements, it is necessary to integrate equation (10.2.1) over the neutron spectrum, a very difficult task often requiring a knowledge of neutron strength functions out to $\ell = 5$ or 6. The writer does not believe that the state of the art at present allows any conclusions about the 11 kev cross sections to be based on such measurements which will give values superior to those obtained from the simpler theoretical procedure just outlined in this chapter.

10.2.1 Supplementary Notes: Average neutron-capture cross sections

The theory of these cross sections can now be improved in three ways. Newton's level-spacing formula has been improved by the writer [110]. With the consequent improvement in the elimination of level-spacing effects in total nuclear radiation widths, it has been possible to discern mass number-dependent fine structure in the total radiation widths which has been attributed to the enhancement of the admixtures of single-particle wave functions in the initial and final states connected by the transitions [111]. It has also become possible to find shell- and deformation-dependent properties in the s-wave neutron strength functions [111].

Chapter 11

Neutron Capture on a Slow Time Scale

We have seen that neutrons can be produced in stellar interiors at temperatures in the vicinity of 100×10^6 °K, and that after being slowed down to thermal equilibrium with their surroundings, they are captured by the nuclei present in proportion to their abundances and capture cross sections. We have also seen the nature of the initial abundance distribution of the nuclei and their neutron-capture cross sections. We shall now consider the changes which take place in the abundances of the nuclei as neutron capture takes place.

11.1 Solution of the abundance equations

If the abundance of the nuclide with mass number A is $N(A)$ and the number of neutrons injected into the abundance unit is $N(n)$, then the rate of change of the abundances is

$$\frac{dN(A)}{dN(n)} = \frac{N(A-1)\sigma(A-1) - N(A)\sigma(A)}{\sum N\sigma}. \tag{11.1.1}$$

In solving these equations, we neglect, for the time being, the effect of neutron absorption by N^{14} and consider that all neutrons are captured by nuclei with mass numbers 20 or higher. The initial abundances $N(A)$ are shown in Figure 9.4.1 and the cross sections in Figure 10.2.3. We are interested in the abundances which are produced as a function of

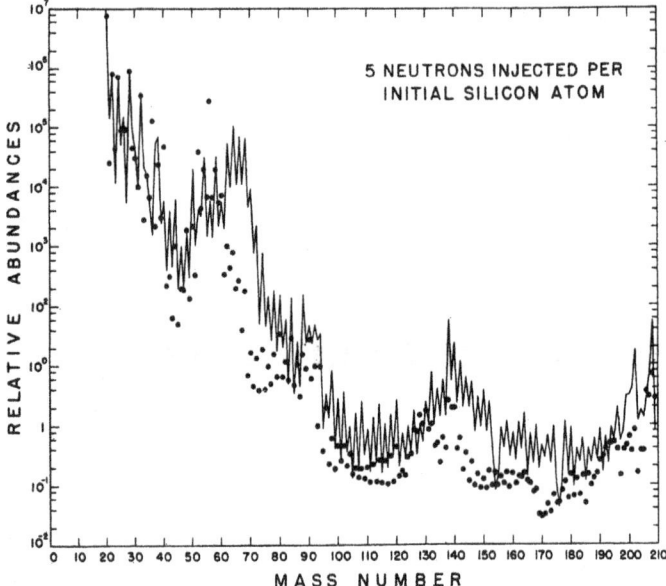

Figure 11.1.1: The abundances produced by the capture of five neutrons injected per initial silicon atom. The ordinates correspond to an initial abundance of 10^6 silicon atoms. The initial abundances are indicated by solid dots.

the total number of neutrons injected, $N(n)$.

The abundance equations were integrated with the aid of an electronic computer. All nuclei were initially considered to lie on the capture path except certain ones with large abundances such as Fe^{54} and Ni^{58}, whose rate of feeding into the capture path was followed independently.

The abundances produced by the capture of 5, 10, 15, 20, 30, 50, 80, and 125 neutrons (injected per initial silicon atom) are shown in Figures 11.1.1 through 11.1.8.

The neutrons are initially captured mostly by the nuclei in the iron peak, centered about $A = 56$. As the capture progresses, a few nuclei capture many neutrons and are moved up to high mass numbers. This "tail" rapidly grows until the abundances of the heavy nuclei become much larger than the initial values. The ratios of the evolved to initial abundances of these nuclei will be called their overabundance factors.

After the injection of 5 neutrons per initial silicon atom, the nuclei around $A = 70$ have become quite overabundant, but heavier nuclei have only small overabundance

110

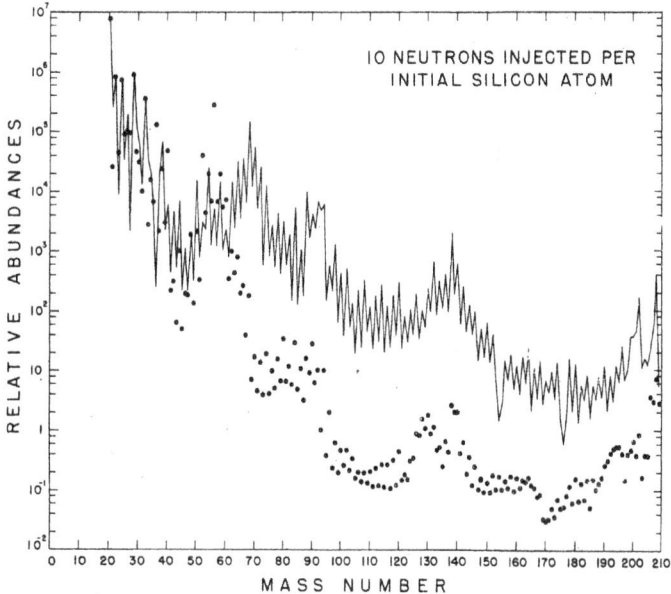

Figure 11.1.2: The abundances produced by the capture of 10 neutrons injected per initial silicon atom. The ordinates correspond to an initial abundance of 10^6 silicon atoms. The initial abundances are indicated by solid dots.

factors. By 10 neutrons, the heavy nuclei have become overabundant by factors of the order of 100. The overabundance factors continue to grow as more neutrons are added until by 50 neutrons, factors of many thousand are reached. For still more neutrons, the overabundance factors decline. They reach a minimum for 125 neutrons. At this point the nuclei originally in the iron peak have been transformed into lead and bismuth, and the overabundance factors are maintained by capture in lighter nuclei.

The reason that the neutron capture does not build up nuclei heavier than lead and bismuth is that above bismuth the capture path enters a region where the nuclei have extremely short half-lives for alpha-particle emission. Hence any nuclei which get that far immediately decay back to lead isotopes.

It should be noticed that at all times during the neutron capture the general level of abundances below $A = 140$ is between a factor of two and a factor of ten higher than the general level above $A = 140$. This is a very significant point which we shall refer to when analyzing the cosmic abundances of the nuclides.

111

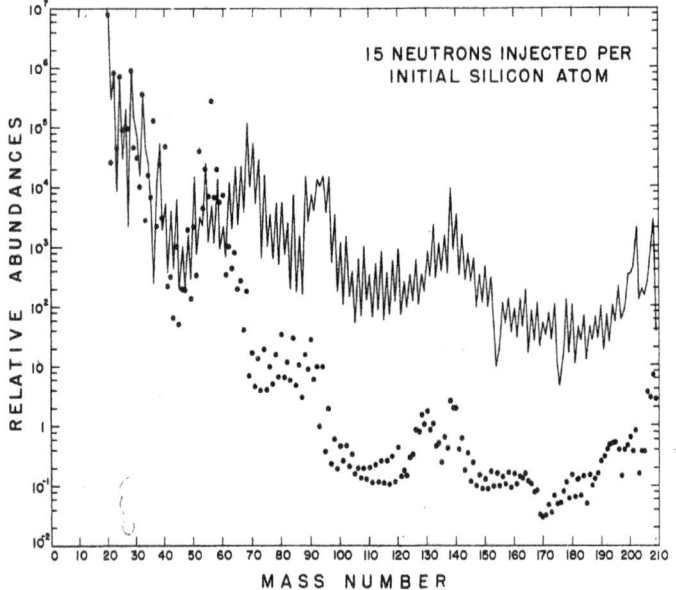

Figure 11.1.3: The abundances produced by the capture of 15 neutrons injected per initial silicon atom. The ordinates correspond to an initial abundance of 10^6 silicon atoms. The initial abundances are indicated by solid dots.

To show the nature of the abundance increases in a little more detail, the overabundance factors of a few selected nuclides are plotted as functions of neutron number in Figures 11.1.9 and 11.1.10. The peak overabundance factors may be expected to have probable errors of factors of two or three owing to probable errors which exist in the formulas from which the cross sections were calculated.

One other significant piece of information which can be obtained from these integrations is the total-absorption cross section of the heavy elements for neutrons as a function of the number of neutrons injected. This is shown in Figure 11.1.11. Owing to the errors in the cross sections, the ordinates in this figure should be approximately doubled.

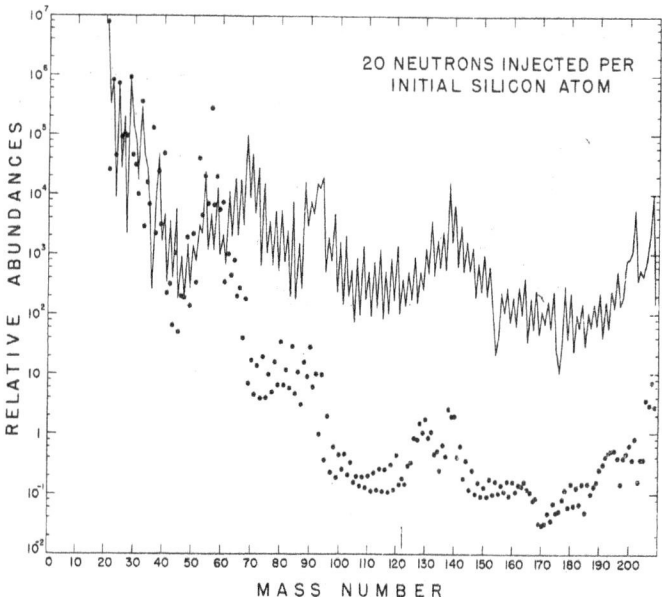

Figure 11.1.4: The abundances produced by the capture of 20 neutrons injected per initial silicon atom. The ordinates correspond to an initial abundance of 10^6 silicon atoms. The initial abundances are indicated by solid dots.

11.2   Possible conditions for heavy-element synthesis

In the earlier discussion it was postulated that the onset of helium thermonuclear reactions causes the core of a red giant star to expand, and that the large amount of energy generation accompanying the production of some C^{12} from the helium causes some hydrogen from the envelope to be mixed into the core. This hydrogen would react with the fresh C^{12} to produce other carbon-cycle isotopes. We see from Figure 8.5.1 that if only a little hydrogen is mixed with C^{12}, the main product formed will be C^{13}, but if a lot of hydrogen is mixed with C^{12}, then we will get carbon-cycle equilibrium abundances in which most of the material will be N^{14}.

Now N^{14} is a very efficient absorber of neutrons through the reaction $N^{14}(n, p)C^{14}$, which is about 25 times as probable as the reaction $N^{14}(n, \gamma)N^{15}$. The C^{14} produced has a half-life of 5600 years, which is comparable to the time postulated previously for neutron production to take place in a stellar interior by the $C^{13}(\alpha, n)O^{16}$ reaction. Thus N^{14}

113

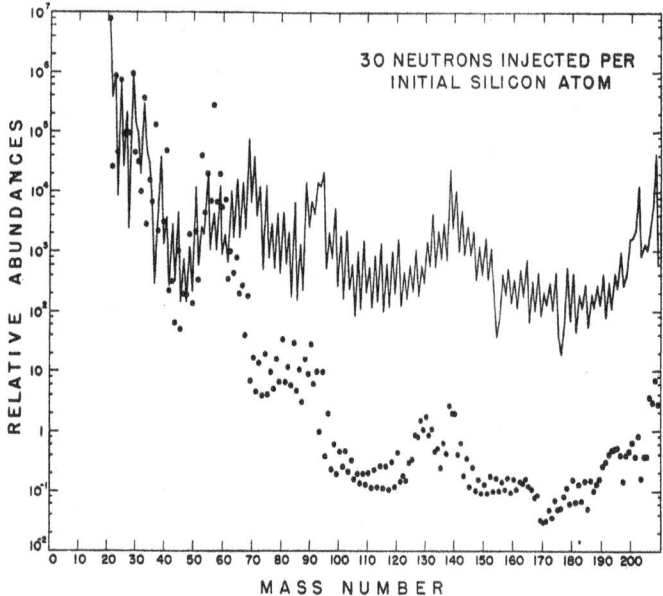

Figure 11.1.5: The abundances produced by the capture of 30 neutrons injected per initial silicon atom. The ordinates correspond to an initial abundance of 10^6 silicon atoms. The initial abundances are indicated by solid dots.

competes very efficiently with the heavy elements for the neutrons which are produced. It is necessary to take detailed account of the absorption by N^{14} in any consideration of heavy-element synthesis where the neutrons are produced by the $C^{13}(\alpha, n)O^{16}$ reaction. Marion & Fowler [42] have pointed out that if neutron production takes place by the $Ne^{21}(\alpha, n)Mg^{24}$ reaction, then the N^{14} may have been destroyed by the $N^{14}(\alpha, \gamma)F^{18}$ reaction.

Some calculations have been carried out to determine whether heavy-element synthesis can take place in the presence of N^{14}. We have seen that more than about 8 neutrons per initial silicon atom must be captured by the heavy elements if large overabundance factors are to be produced. The calculation took the following form:

(1) It was assumed that in the expansion of the stellar core, the helium thermonuclear reactions create X atoms of C^{12} per initial He^4 atom. Calculations were carried out for several values of X ranging from 0.01 to 0.12.

114

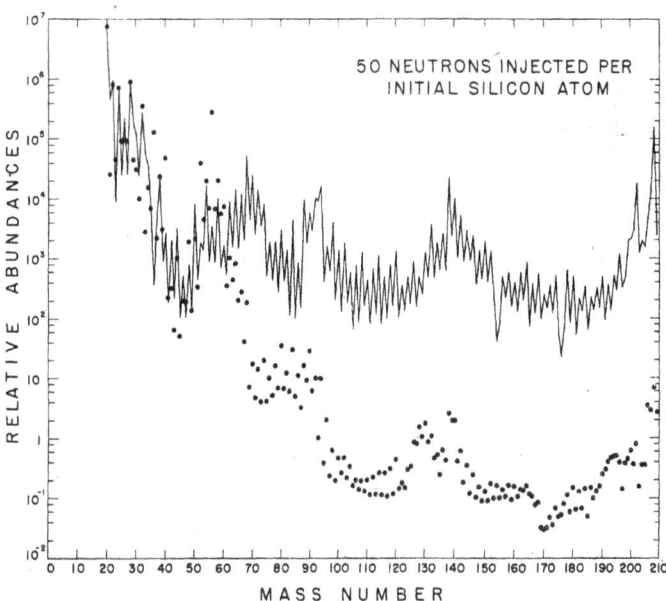

Figure 11.1.6: The abundances produced by the capture of 50 neutrons injected per initial silicon atom. The ordinates correspond to an initial abundance of 10^6 silicon atoms. The initial abundances are indicated by solid dots.

(2) It was assumed that some hydrogen was mixed into this core material, the ratio of the number of admixed protons per C^{12} atom being treated as a variable quantity in the calculations. This hydrogen was allowed to react with the C^{12} nuclei, producing other carbon-cycle isotopes in a manner similar to that shown in Figure 8.5.1. Calculations were carried out for both the old value of the $N^{14}(p,\gamma)O^{15}$ reaction rate and for this reaction rate reduced by a factor of ten.

(3) The C^{13} left over after the exhaustion of hydrogen in the above mixture was allowed to react with helium, producing neutrons. The competition for the neutrons by the N^{14} present and by the heavy elements was taken into account, using the relation shown in Figure 11.1.11 for the variation in the heavy-element absorption cross section with the number of neutrons captured by them. It should be noted that the capture of neutrons by heavy elements will have been underestimated by this process, owing to the errors of about a factor two in the cross sections, and hence

115

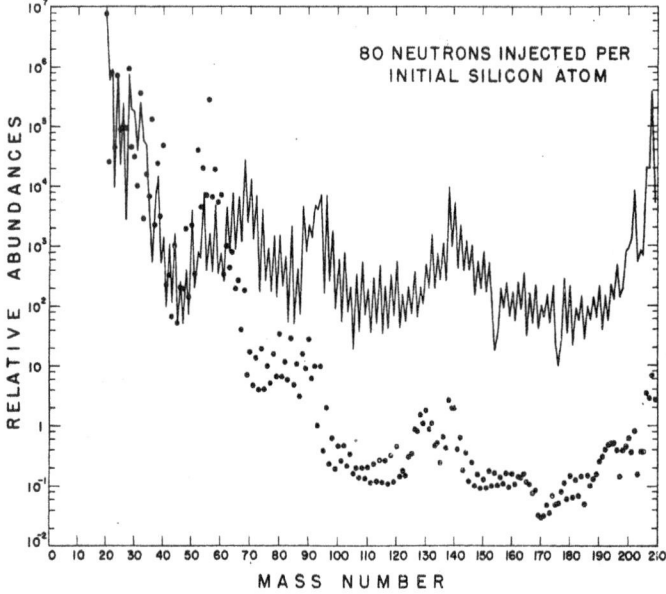

Figure 11.1.7: The abundances produced by the capture of 80 neutrons injected per initial silicon atom. The ordinates correspond to an initial abundance of 10^6 silicon atoms. The initial abundances are indicated by solid dots.

the present computation is a conservative one.

Several subsidiary conditions had to be taken into account at this stage. The $N^{14}(n, p)C^{14}$ reaction produces protons. These interact with the carbon-cycle nuclei present to transform their abundances, and it was necessary to follow these reactions in order to determine the amount of C^{13} regenerated. Calculations were carried out for two extreme cases: the assumption that the C^{14} produced immediately decayed to N^{14} (i.e., in a time short compared to the time of the neutron production), and the assumption that it did not decay to N^{14} at all. In the former case the amount of N^{14} neutron absorber stays constant or even increases owing to the effects of the protons produced from it. In the latter case the amount of N^{14} decreases but one must take into account capture protons by the $C^{14}(p, \gamma)N^{15}$ reaction. Calculations were made for two assumptions about this proton capture: that it took place in the tails of known resonances in the N^{14} nucleus and that it took place through a thermal resonance. In all cases the calculation

116

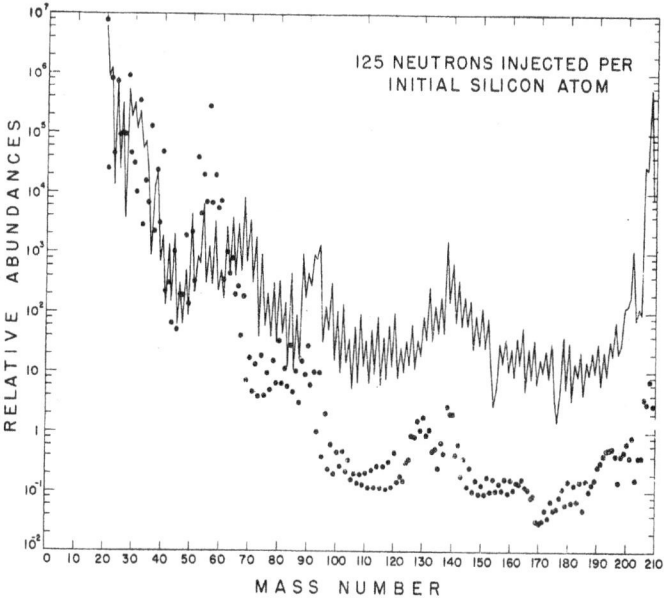

Figure 11.1.8: The abundances produced by the capture of 125 neutrons injected per initial silicon atom. The ordinates correspond to an initial abundance of 10^6 silicon atoms. The initial abundances are indicated by solid dots.

was continued until the C^{13} was completely exhausted by reactions with helium, and the total number of neutrons captured by the heavy elements had been obtained.

The above calculations were carried out for the Suess-Urey abundances of Figure 9.4.1, which may be regarded as solar abundances. The situation which would exist in Population II stars was also studied by performing some of these calculations with a reduction by a factor of ten in the abundances of the elements relative to hydrogen. The results are shown in Figures 11.2.1 through 11.2.3. In Figure 11.2.1 it may be seen that even though C^{14} is assumed to decay immediately to N^{14}, the heavy elements will be produced with large overabundances if $X = 0.02$ and if 0.05 to 0.2 protons are admixed per C^{12} atom. In Figure 11.2.2 we see that we have a somewhat wider range of acceptable conditions where the C^{14} is assumed not to decay to N^{14}. Finally, we see in Figure 11.2.3 that large overabundances of heavy elements can be produced in Population II stars over a wide range of mixing conditions.

117

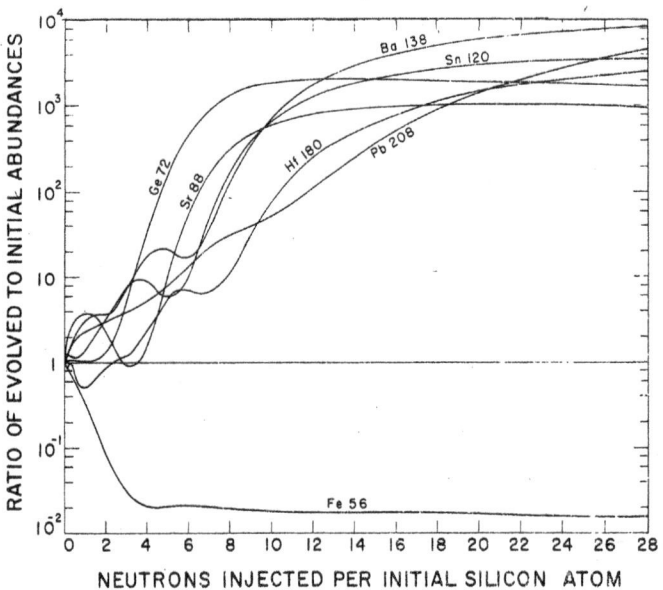

Figure 11.1.9: Overabundance factors (ratios of evolved to initial abundance) for some selected nuclides plotted as a function of the number of neutrons up to 28 neutrons injected per initial silicon atom.

If in the process where hydrogen is mixed into the core the hydrogen does not reach the center, then a gradient in the proton admixture ratio must be set up. The above results indicate that large overabundances of heavy elements will then be produced in a shell of undetermined thickness.

There are still further conditions under which the heavy elements can be synthesized. Let us consider the equilibrium abundances in the carbon cycle at very high temperatures. These are shown in Figure 11.2.4 and were calculated using the old value of the $N^{14}(p, \gamma)O^{15}$ reaction rate. It may be seen that the ratios of the carbon isotopes relative to N^{14} increase with the temperature until temperatures of the order of 100×10^6 °K are reached. At this point the mean life of the C^{12}, C^{13}, and N^{14} nuclei against proton capture becomes less than 10 minutes – the beta half-life of N^{13}. Hence the most abundant member of the carbon cycle becomes N^{13} with O^{15} the second most abundant member. This state of affairs continues until temperatures in the vicinity of 300×10^6 °K are reached (it should be noted that, essentially, a constant rate of energy generation is

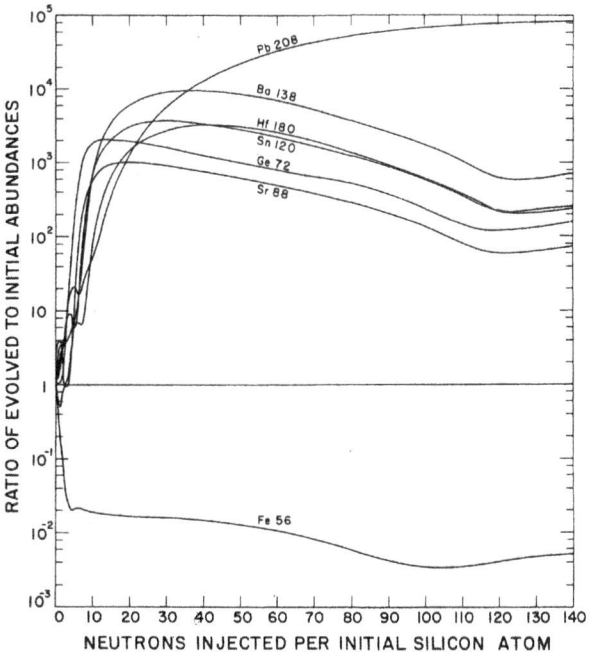

Figure 11.1.10: Overabundance factors (ratios of evolved to initial abundance) for some selected nuclides plotted as a function of the number of neutrons up to 140 neutrons injected per initial silicon atom.

obtained from the carbon cycle in this temperature range). At 300×10^6 °K the N^{13} is destroyed by the reaction

$$N^{13}(p, \gamma)O^{14}. \tag{11.2.1}$$

The reason that this reaction is so slow is that the thermonuclear energy region in O^{14} is at an excitation energy of 4.6 Mev, but the first excited state is at 6 Mev. Hence the reaction can go only in the tails of distant levels. At a temperature of 400×10^6 °K the reaction

$$N^{13}(\gamma, p)C^{12} \tag{11.2.2}$$

starts to become important. This soon prevents the carbon cycle going beyond C^{12}, which becomes the most abundant nucleus.

We see that from 100 to 200×10^6 °K, the abundance of N^{13} is enormously greater than those of N^{14} and O^{14} combined. Hence if the hydrogen swept into the stellar core

119

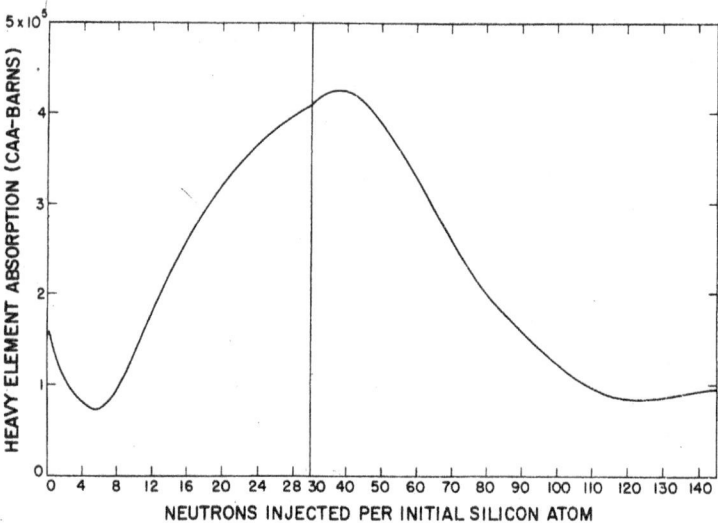

Figure 11.1.11: Total-absorption cross section of the heavy elements for neutrons as a function of the number of neutrons captured. This is the cross section summed over all nuclides in a cosmic abundance unit initially containing 10^6 atoms of silicon, but excluding N^{14}.

is exhausted at temperatures above 100×10^6 °K it will leave behind nearly pure N^{13} plus O^{15}, which after beta decay will become C^{13} and N^{15}. If such material is then allowed to react with helium, the neutrons produced all will be captured by the heavy elements. If this case can take place, there is a very wide range of conditions which will synthesize the heavy elements — one need only have enough parent material to produce the neutrons. In the sun there are 11 carbon plus nitrogen atoms per silicon atom and 25 oxygen atoms per silicon atom. Therefore this is a mechanism which might synthesize the heavy elements in very hot stars of solar composition, even without the assistance of carbon production in a stellar core.

11.3 Stars with abnormal heavy-element abundances

We now consider stars which have been observed to have large abundances of heavy elements. Bidelman [112] remarks that there is a general strengthening of the lines of the heavy elements in stars with S spectra. However, so far there is very little quantitative

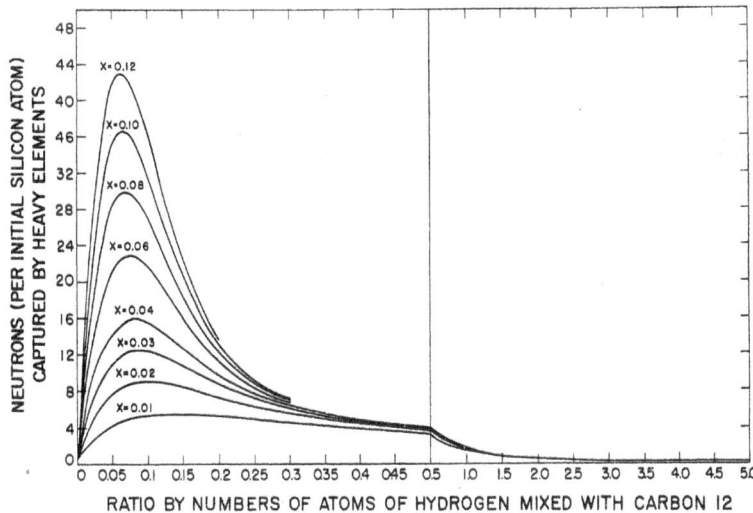

Figure 11.2.1: Number of neutrons captured by heavy elements under the assumption that C^{14} immediately decays to N^{14}, using old values for the $N^{14}(p, \gamma)O^{15}$ reaction rate. The results are hardly effected except in the right half of the figure if the $N^{14}(p, \gamma)O^{15}$ reaction rate is decreased by a factor of 10.

data about these stars. Buscombe & Merrill [113] found that the abundance of zirconium was increased by factors of 10 and 60 in two S stars as compared to a normal M star. The star with the higher abundance is R Andromeda, and Greenstein now has a program in progress to determine many abundances in this star. Fowler, Burbidge & Burbidge [93] estimate that the rare earths in general are increased in abundance in S spectra by factors of 10 to 100, but the writer believes that a current consensus of opinion would favor the lower of these limits as being closer to the truth. These facts suggest that S spectra may be produced by mixing of the order of a percent of neutron-evolved material into the surface layers of giant stars. These stars also contain an increased carbon content, which is consistent with the idea that neutron production has been associated with carbon production in the stellar core.

But the most striking abundance anomaly in the S stars is the presence of the unstable element technetium [114]. The isotope of technetium with longest known half-life is Tc^{99}, a member of the neutron-capture path considered to be stable since its half-life is

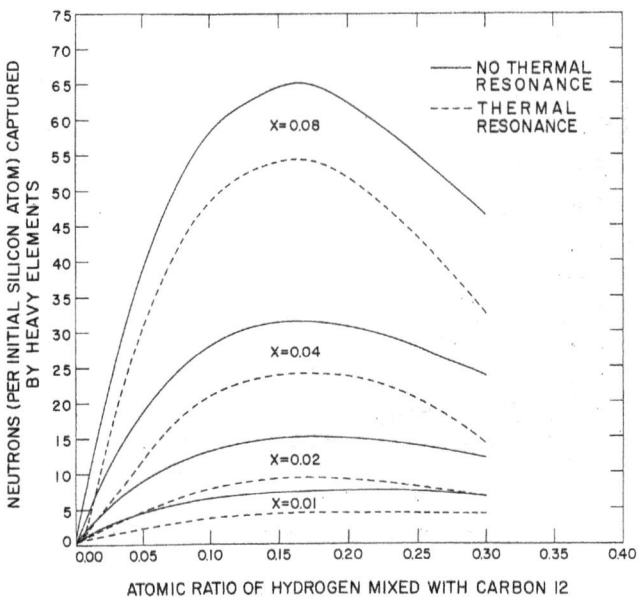

Figure 11.2.2: Number of neutrons captured by heavy elements under the assumption that C^{14} does not decay to N^{14}. The dotted curves correspond to the assumption of a thermal resonance in the $C^{14}(p, \gamma)N^{15}$ reaction.

210,000 years. Greenstein [91] estimates that the ratio of abundances of Tc and Fe in R Andromeda is about 5×10^{-5}. This is of the order of magnitude to be expected in a neutron-capture synthesis followed by a mixing of a small amount of the products into the stellar surface layers. Evidently, not many Tc^{99} half-lives have passed since some of the surface material in R Andromeda was exposed to neutrons. At the same time it should be noted that the Nb lines are abnormally strong in this star. The only stable isotope of this element is Nb^{93}, which does not lie on the neutron-capture path. Instead, the path contains Zr^{93} which has a half-life of 9×10^5 years, decaying to Nb^{93}. Evidently, some of the surface material in R Andromeda stopped being exposed to neutrons long enough ago for an appreciable fraction of the Zr^{93} atoms to have decayed to Nb^{93}.

Merrill [115] has observed technetium in two stars with N spectra but has failed to find it in two stars with R spectra.

Stars with higher surface temperatures which appear to have compositions similar to those of the S stars are stars classed as CH or Ba II stars. Burbidge & Burbidge [116]

122

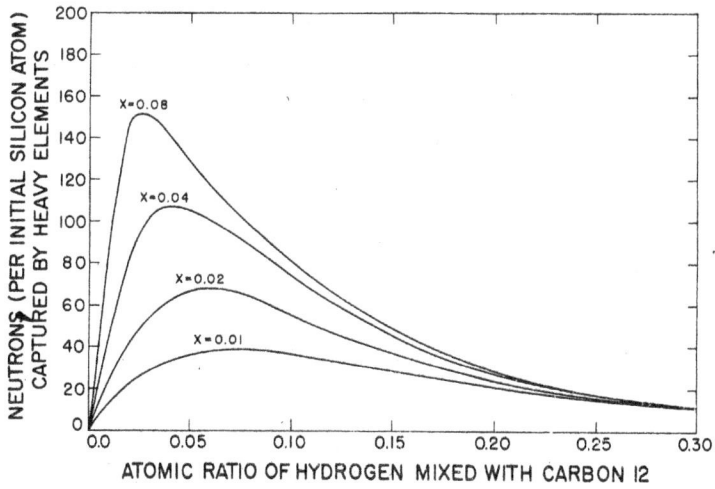

Figure 11.2.3: Number of neutrons captured by heavy elements in stars with 10 percent of the solar abundances of heavy elements. The C^{14} was assumed to decay immediately to N^{14}.

have obtained quantitative abundances in the Ba II star HD 46407. They found that 16 elements in the range up to and including germanium have essentially normal abundances in this star. Most of the heavier elements are overabundant by factors greater than 4 and (in the case of Pr) as high as 28. The heavier element group includes 17 elements. These observations are strongly suggestive that a little less than one percent of neutron-evolved material has been mixed into the surface layers of this star. No technetium has been found, indicating that neutron production ceased affecting the surface layers a long time ago.

It should be mentioned that certain peculiar A and P stars contain large overabundances of the heavy elements. However, the overabundance factors disagree quantitatively with the hypothesis that they have been produced by neutron capture on a slow time scale, and an alternate hypothesis will be discussed later by which these overabundances have been produced by nuclear reactions in the stellar surface.

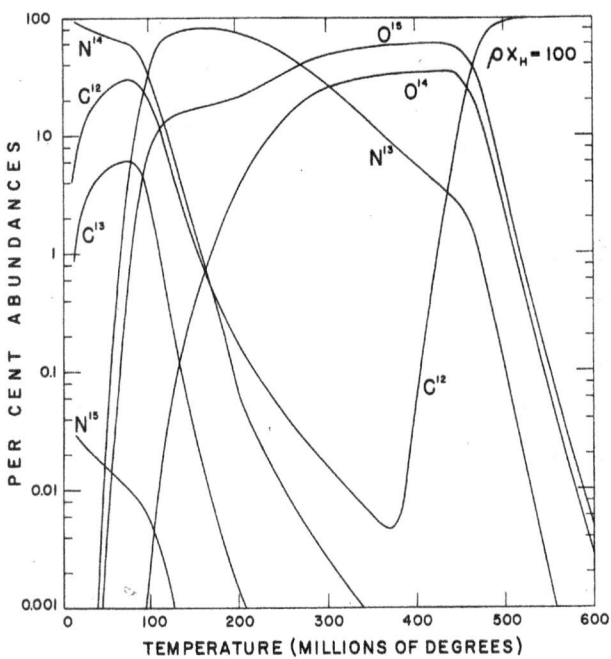

Figure 11.2.4: The equilibrium abundances of nuclides in the carbon cycle at elevated temperatures.

Chapter 12

Heavy-Ion Thermonuclear Reactions

We have considered in some detail the nuclear reactions which can go on in stellar interiors up to the point where the helium has been exhausted. Beyond this stage our discussion must become very qualitative, both because we have no stellar models to guide us, and because the thermonuclear reactions which can take place have not yet been analyzed in detail. Apparently, some of the stars can evolve to very advanced stages in which the central temperatures reach 10^9 °K. At these higher temperatures new thermonuclear reactions set in which involve reactions between the products of the helium thermonuclear reactions, C^{12}, O^{16}, and Ne^{20}. Because such particles are called heavy ions when accelerated in the laboratory, we shall call the reactions between them heavy-ion thermonuclear reactions.

Since the production of a high temperature in the region of 10^9 °K may be accompanied by the attainment of very large densities in stellar interiors, the onset of heavy-ion thermonuclear reactions may induce a new instability in the degenerate gas of the stellar core. This instability may cause further internal mixing to take place in the star.

Heavy-ion thermonuclear reactions have been briefly discussed by Hoyle [77] and by Nakagawa et al. [78]. Hoyle has attempted to follow some of the details of the reactions, but his discussion is grossly oversimplified and incorrect in many of the details.

In order to get a feeling for the sort of things that can happen in heavy-ion thermonuclear reactions, we reproduce a list of the exothermic reactions given by Nakagawa et al. in Table 12.1. Two things are immediately obvious from an inspection of this table. The first is that heavy-ion thermonuclear reactions take place at very high excitation energies in compound nuclei of intermediate mass. The level densities in the thermonuclear energy regions are, therefore, very large, and one can calculate reaction rates assuming the nonresonant formula taken with a nuclear cross section which is an average over nuclear resonances.

The second noticeable fact is that there are many reactions in which particles are emitted. The particles of particular importance are neutrons, protons, and alpha-particles.

The neutrons released at lower temperatures will be captured by the heavier elements, giving additional neutron-capture products on the slow time scale.

The protons produced cannot add to C^{12}, O^{16}, or Ne^{20} because of the high rates of proton removal by photodisintegration of N^{13}, F^{17}, and Na^{21}, but they will be absorbed by the other lightest nuclei present. The alpha-particles will also react with the lightest particles present.

Reactions between O^{16} and O^{16} do not start until a temperature of about 1.35×10^9 °K is reached and those of the Ne^{20} with itself start at an even higher temperature. However, these reactions cannot be considered in the absence of accompanying photodisintegrations, to be discussed later.

The general result of the heavy-ion thermonuclear reactions is to start building nuclei of intermediate weight beyond neon, and to give further contributions to neutron capture on a slow time scale.

Table 12.1: Exothermic Heavy-Ion Reactions

Bombarding Particles	Products	Energy Release (Mev)
$C^{12} + C^{12}$	$\gamma + Mg^{24}$	13.95
	$p + Na^{23}$	2.25
	$\alpha + Ne^{20}$	4.62
$C^{12} + O^{16}$	$\gamma + Si^{28}$	16.8
	$p + Al^{27}$	5.21
	$\alpha + Mg^{24}$	6.80
$C^{12} + Ne^{20}$	$\gamma + S^{32}$	19.0
	$n + S^{31}(\beta^+\nu)P^{31}$	4.2
	$p + P^{31}$	10.1
	$\alpha + Si^{28}$	12.0
	$Be^8 + Mg^{24}$	2.0
	$O^{16} + O^{16}$	2.4
$O^{16} + O^{16}$	$\gamma + S^{32}$	16.6
	$n + S^{31}(\beta^+\nu)P^{31}$	1.8
	$p + P^{31}$	7.7
	$\alpha + Si^{28}$	9.7
$O^{16} + Ne^{20}$	$\gamma + A^{36}$ †	18.5
	$n + A^{35}(\beta^+\nu)Cl^{35}$	3.8
	$p + Cl^{35}$	10.0
	$\alpha + S^{32}$	11.8
	$Li^5 + P^{31}$	1.2
	$Be^8 + Si^{28}$	4.8
	$C^{12} + Mg^{24}$	2.2
$Ne^{20} + Ne^{20}$	$\gamma + Ca^{40}$	20.9
	$n + Ca^{39}(\beta^+\nu)K^{39}$	5.0
	$p + K^{39}$	12.5
	$d + K^{38}$	1.5
	$He^3 + A^{37}(e,\nu)Cl^{37}$	2.0
	$\alpha + A^{36}$	13.7
	$Li^5 + Cl^{35}$	3.4
	$Be^8 + S^{32}$	7.0
	$C^{12} + Si^{28}$	7.4
	$O^{16} + Mg^{24}$	4.6

†Editor's Note: The International Commission on Atomic Weights adpoted a change of the symbol of argon from "A" to "Ar" in 1957.

Chapter 13

Photonuclear Reactions on a Slow Time Scale

Photodisintegration reactions become important in the same range of temperature at which heavy-ion thermonuclear reactions take place, and it is necessary to consider the two kinds of reactions together. Hoyle [77] has indicated the basic method of computing photodisintegration rates.

Although we have so far encountered only one case in which we have a statistical equilibrium set up between two nuclei ($Be^8 \rightleftharpoons 2He^4$), nevertheless we will always have in stellar interiors an equilibrium set up between capture reactions and their photodisintegration inverses. This is because, as we shall see later, the Planck distribution of radiation is maintained to an excellent degree of approximation. Usually, however, the equilibrium favors one reaction rate enormously more than its inverse.

If we have equilibrium set up between the number densities n_1 and n_2 of two constituents and n_{12} of their capture compound, then from statistical mechanics we know the relative amounts of each formed:

$$\frac{n_1 n_2}{n_{12}} = \left(\frac{2\pi m_1 m_2 k}{m_1 + m_2} \right)^{3/2} \frac{T^{3/2}}{h^3} \frac{\omega_1 \omega_2}{\omega_{12}} \, e^{\chi/kT}, \qquad (13.0.1)$$

where ω_1, ω_2, and ω_{12} are the statistical weights of the partciles and χ is the energy of the ground state of the compound system above the sum of the masses of the constituent particles.

Now the probability of photodisintegration per unit time can be set equal to λn_{12}. On the other hand, the probability of formation per unit time is

$$r = n_1 n_2 \langle \sigma v \rangle = q n_1 n_2. \tag{13.0.2}$$

In equilibrium the disintegration and formation rates are equal:

$$\lambda n_{12} = q n_1 n_2,$$

$$\text{or } \lambda = q \frac{n_1 n_2}{n_{12}} \tag{13.0.3}$$

$$= 1.89 \times 10^{35} \left(\frac{A_1 A_2}{A_1 + A_2} \right)^{3/2} \frac{\omega_1 \omega_2}{\omega_{12}} q T^{3/2} \, \mathrm{e}^{\chi/kT}.$$

Hoyle [77] has pointed out that at a temperature of 0.8×10^9 °K, the following reactions set in:

$$\mathrm{Ne}^{20}(\gamma, \alpha)\mathrm{O}^{16}. \tag{13.0.4}$$

At 1.3×10^9 °K we also get the reaction

$$\mathrm{O}^{16}(\gamma, \alpha)\mathrm{C}^{12}. \tag{13.0.5}$$

These reactions strongly complicate the situation in which we have heavy-ion thermonuclear reactions taking place.

Let us consider the reactions which take place which destroy the heavy nuclei which have been built up by neutron capture on a slow time scale. Since photonuclear reactions are primarily sensitive to particle binding energies, we can get sufficiently accurate results for heavy nuclei, if we assume they all have a neutron-capture cross section of about 0.4 barns at 11 kev, and that this cross section varies as $1/v$. This means that q is constant, equal to

$$q = 5.8 \times 10^{-17}.$$

Hence

$$\lambda \approx 2.2 \times 10^{13} T^{3/2} \exp \left(\frac{-0.0116 B_n}{T} \right) \, \mathrm{s}^{-1}, \tag{13.0.6}$$

129

Table 13.1: Photodisintegration rates (s^{-1}) as a function of neutron binding energy (Mev).

B_n / T	5	6	7	8	9	10
1000	4.4×10^{-8}	4.2×10^{-13}	4.0×10^{-18}	3.5×10^{-23}	2.8×10^{-28}	2.8×10^{-33}
1200	9.3×10^{-4}	5.9×10^{-8}	3.8×10^{-12}	2.4×10^{-16}	1.5×10^{-20}	9.3×10^{-25}
1400	1.2×10^{0}	2.9×10^{-4}	7.3×10^{-8}	1.8×10^{-11}	2.1×10^{-15}	1.2×10^{-18}
1600	2.5×10^{2}	1.7×10^{-1}	1.4×10^{-4}	2.2×10^{-8}	5.7×10^{-11}	4.9×10^{-14}
1800	1.7×10^{4}	2.6×10^{1}	4.3×10^{-2}	6.7×10^{-5}	1.1×10^{-7}	1.7×10^{-10}
2000	5.0×10^{5}	1.5×10^{3}	4.7×10^{0}	1.4×10^{-2}	4.5×10^{-5}	1.3×10^{-7}
2200		4.0×10^{4}	2.2×10^{2}	1.1×10^{0}	5.6×10^{-3}	2.9×10^{-5}
2400			5.2×10^{3}	4.2×10^{1}	3.3×10^{-1}	2.6×10^{-3}
2600			7.8×10^{4}	9.1×10^{2}	1.1×10^{1}	1.2×10^{-1}
2800				1.3×10^{4}	2.1×10^{2}	3.3×10^{0}
3000				1.4×10^{5}	2.8×10^{3}	5.8×10^{1}

where T is measured in units of 10^6 °K, and B_n is the neutron binding energy in electron volts. A list of photodisintegraion rates as a function of neutron binding energy is given in Table 13.1.

Now let us supose we have a stellar condition in which the temperature at the center is rising about ten percent per 10^7 years. This will give us a rough time scale for photodisintegration effects, and we will be interested in reactions which have photodisintegration rates of $\sim 10^{-14}$ s^{-1}. For substances with neutron binding energies of 8 Mev, these rates are obtained at $T \sim 1300 \times 10^6$ °K. Hence neutrons will be stripped out of heavy nuclei at temperatures in this range. Let us suppose we start with a heavy nucleus on the main neutron-capture path with even numbers of neutrons and protons. After a neutron is removed, the binding energy of a neutron in the product nucleus is about 1 Mev less than in the starting nucleus. Hence the photodisintegration rate for the product nucleus will be $\sim 10^{-10}$ s^{-1} corresponding to a mean life ~ 100 years. However, many product nuclei will be unstable to positron emission or electron capture with mean lives of less than 100 years, and sometimes to negative beta decay. Therefore, we will have a series of photoneutron reactions interspersed with beta transformations, which will gradually break down the heavy nuclei into the region of the iron peak. This process will be called photodisintegration on the slow time scale.

Several nuclei, particularly certain even-even isobars, will be formed in these photo-disintegration reactions which are not formed by neutron-capture reactions. However, not all neutron-deficient stable isobars can be formed in this way.

Chapter 14

Nuclear Reactions in Statistical Equilibrium

As temperatures and densities still higher than those so far considered are reached, a very large number of nuclear reactions become possible and take place at very rapid rates. It is at present hopeless to try to follow these reactions in any detail, but if full statistical equilibrium is attained, then it is possible to calculate the equilibrium abundances of all the nuclei. This has been done by a number of people, but the present discussion will follow that of Hoyle [117], since his paper forms the basis for the supernova model which we shall later discuss.

14.1   The conditions for statistical equilibrium

There are a number of conditions which must be satisfied if statistical equations are to be valid. These are:

(1) Energy must be statistically distributed among the energy states of translation of each type of particle present. This requirement will be satisfied provided all types of particle experience an appreciable number of collisions per second. Under conditions of interest, we will have of the order of 10^{29} particles per cm^3 all with velocities of the order of 10^9 cm/s. The smallest collision cross sections are for the neutrons which have scattering cross sections of the order of 10^{-24} cm^2. The

132

product of these quantities is the number of collisions per second; with a value of 10^{14}, it satisfies our requirement by a large margin.

(2) There must be a thermodynamic equilibrium between matter and radiation. We have already assumed this in calculating photodisintegration rates. Since the energies of the photons emitted in capture reactions are large compared to kT, the Planck distribution function must hold up to values very large compared to kT. To establish this condition, we must show that a sufficiently large interchange of energy takes place between matter and radiation and that it takes place up to energies large compared to kT. We give this requirement the precise definition that the amount of thermal energy of material converted into radiant energy per second per cm^3 in any frequency range ν to $\nu + d\nu$ must be large compared with the equilibrium energy density of radiation given by the Planck formula

$$\frac{8\pi h\nu^3 d\nu}{c^3(\exp(h\nu/kT) - 1)}. \qquad (14.1.1)$$

Under these conditions, detailed balancing will occur, and equal quantities of radiant energy will be converted into thermal energy. We consider $h\nu \gg kT$.

The number of electrons with energy greater than E_0 per cm^3 is given by

$$\frac{2n_e}{\pi^{1/2}(kT)^{3/2}} \int_{E_0}^{\infty} \exp\left(\frac{-E}{kT}\right) E^{1/2} dE, \qquad (14.1.2)$$

where n_e is the number density of electrons. This is valid since condition (1) is satisfied. For $E_0 \gg kT$, we may make the approximation that the above number is:

$$2n_e \left(\frac{E_0}{\pi kT}\right)^{1/2} \exp\left(\frac{-E_0}{kT}\right). \qquad (14.1.3)$$

Now an electron can lose nearly all its energy by bremsstrahlung emission in a collision with a positive ion. If the ion has charge Z, the cross section for this process is about $5 \times 10^{-27} Z^2$. Combining this is (14.1.3), then the number quanta

133

with energy greater than E_0 emitted per cm^3 per second is of the order

$$10^{-26} n_e v_e \left(\sum_{A,Z} Z^2 n_A^Z \right) \left(\frac{E_0}{kT} \right)^{1/2} \exp\left(\frac{-E_0}{kT} \right), \qquad (14.1.4)$$

where v_e is the velocity of electrons with energy E_0, and the summation is taken over all nuclei present in the material with number densities n_A^Z. Now, in thermodynamic equilibrium, the number of quanta with energy greater than E_0 per cm^3 is

$$\frac{8\pi}{c^3} \int_{\nu_0 = E_0/h}^{\infty} \left[\frac{\nu^2 d\nu}{\exp(h\nu/kT) - 1} \right] \approx \frac{8\pi E_0^2 kT}{c^3 h^3} \exp\left(\frac{-E_0}{kT} \right). \qquad (14.1.5)$$

We get an order of magnitude estimate for the time required for equilibrium to be set up by dividing (14.1.5) by (14.1.4). For $A = 4$, $Z = 2$, $2n_e = n_4^2 = 10^{30}$ cm^{-3}, $v_e = 10^{10}$ cm/s, $E_0 = 20$ Mev, and $kT = 0.3$ Mev, this gives a time of the order of 10^{-12} second. Hence our second condition is also satisfied by a large margin.

(3) There must be suitable nuclear reactions connecting any two values of Z and A, and a chain of such reactions must be able to take place with a mean reaction time for all the steps short compared to the time in which we wish equilibrium to be established. It is sufficient to consider only the addition of neutrons and protons to nuclei and their removal from nuclei. One can connect any two pairs of values of Z and A by steps consisting of nucleon emissions and additions.

We have already seen that photoneutron emission rates become extremely fast for nuclei in the ordinary range of neutron binding energies for temperatures of 3×10^9 °K and above. We shall see in Section 16.6 that for nuclei in the rare-earth region, the photoproton emission rates are comparable to photoneutron rates, where the neutron binding energy is 4 Mev higher than the proton binding energy. Hence, the conditions associated with the removal of nucleons are generally met for temperatures above 3×10^9 °K.

The rates of addition of nucleons to nuclei will be large enough only provided there are substantial numbers of free protons and neutrons available in statistical equilibrium. When we consider thermonuclear reaction rates, it turns out that even for heavy nuclei there will be fast enough addition rates for neutrons and protons ($\sim$1 per second) if about one part in 10^4 of the number of nucleons in the assembly are available as free neutrons and as free protons, at temperatures greater than 3×10^9 °K. We can see immediately that under conditions of interest, if such a nucleon density does not exist initially, it will rapidly be established by photodisintegration reactions.

(4) Equilibrium between protons and neutrons requires a sufficiently rapid beta transformation to take place. Not only must we have short beta half-lives present, but they must exist in nuclei with large equilibrium abundances. This requires that we have some knowledge of equilibrium abundances, which we will obtain in the next section. The question of fast beta transformations is intimately linked to the rates of the Urca process, also to be discussed in Section 14.3. We shall see that the interchange between neutrons and protons is the slowest of our necessary conditions to be satisfied; under conditions of interest, it may take several hours to establish full equilibrium between neutrons and protons. However, this time becomes much shorter at very high temperatures and densities (5×10^9 °K).

14.2   Equilibrium abundances of the nuclides

The equations of statistical equilibrium can be derived in a number of ways [13]. The abundance of a nuclide n_A^Z is best expressed in terms of n_p, n_n, and T, the number densities of protons and neutrons and the temperature, respectively [117]. The result is:

$$
\begin{aligned}
\ln n_A^Z =&\, Z \ln n_p + (A + Z) \ln n_n + (A - 1) \ln \frac{h^3}{(2\pi kT)^{3/2}} - \frac{Q_A^Z}{kT} \\
&- \frac{3}{2} Z \ln m_p - \frac{3}{2}(A - Z) \ln m_n + \frac{3}{2} \ln m_A^Z + \ln(2I + 2),
\end{aligned}
\tag{14.2.1}
$$

$$
\text{where } Q_A^Z = c^2 \left[m_A^Z - Z m_p - (A - Z) m_n \right], \tag{14.2.2}
$$

and where m_p, m_n, and m_A^Z are the masses of the proton, the neutron, and the nuclide Z, A, respectively.

We need a further relation to make the problem determinate for a given set of conditions of temperature and density. This relation must be one between the number

135

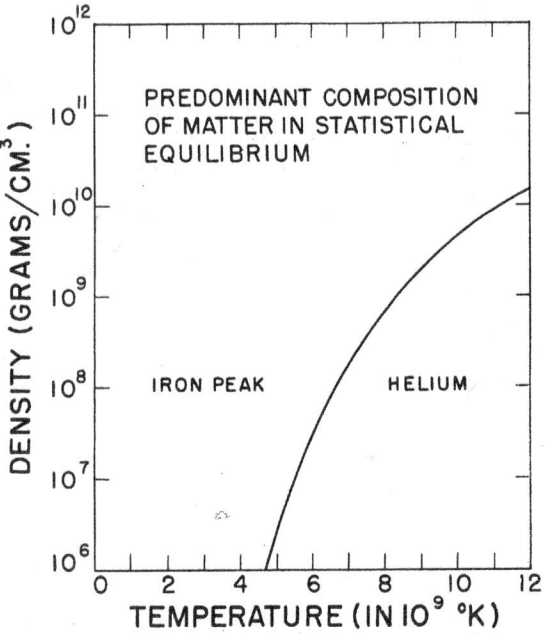

Figure 14.2.1: Regions in the temperature-density diagram in which the material is mostly in the form of the iron peak or of He^4.

densities of the neutrons, protons, and electrons. We can relate these through the beta processes that keep the protons in statistical equilibrium with the neutrons, but it is necessary to use relativistic and degenerate relations for the electrons at high densities. Hoyle finds for high densities:

$$\text{if} \quad x = \left(\frac{n_e}{5.87 \times 10^{29}}\right)^{1/3},$$

$$\text{and} \quad y = (1 + x^2)^{1/2} - 0.51, \tag{14.2.3}$$

$$\text{then} \quad \log_{10}\left(\frac{n_n}{n_p}\right) = \frac{2570y}{T},$$

where T is in units of 10^6 °K.

It should be noted that equation (14.2.1) should be used for calculating the equilibrium abundance of every state (with spin I) of the nucleus n_A^Z, the energy of excitation being included in the mass m_A^Z. The total abundance of n_A^Z is then obtained by summing

136

the abundances of all the excited states.

The general nature of the solutions to these equations has been known for many years [13]. Let us first consider temperatures in the vicinity of 4×10^9 °K. Here the most abundant nuclides are those with the greatest average binding energy per nucleon. The most abundant nuclide is Fe^{56}. Hoyle, Fowler and the Burbidges [118] have carried out recently some very accurate calculations which make use of the latest experimental information about the spins and excitation energies of the excited states in a range of nuclei in the vicinity of Fe^{56}. They find that the relative abundances of the isotopes in the elements in the iron peak region are quite well reproduced. However, the relative abundances of the elements generally agree somewhat better with the solar abundance determinations of Goldberg, Müller & Aller [60] than with the abundances of Suess & Urey [102]. Generally speaking, however, the cosmic abundance peak centered about Fe^{56} in Figure 9.4.1 is quite well reproduced by these calculations.

Let us now consider a higher temperature, for example 6×10^9 °K. When one calculates equilibrium abundances in this case, it turns out that nearly all the material is in the form of He^4. There is a subsidiary peak at Fe^{56}, but this is enormously less abundant than He^4. Evidently, raising the temperature has produced a sort of photodisintegration of the iron peak. At still higher temperatures, the abundances of free neutrons and protons grow until most of the material is in the form of free nucleons.

For a given density, it turns out that there is a very narrow region of temperature in which the equilibrium abundance changes from favoring Fe^{56} by a large factor to favoring He^4 by a large factor. Hence, we can set up a sort of phase diagram in a plot of temperature against density. This is shown in Figure 14.2.1 and is plotted from figures given by Hoyle [117]. On one side of the line most, of the nucleons form Fe^{56}; on the other side most nucleons form He^4. We shall see that the key to an understanding of supernova explosions appears to lie in this diagram.

14.2.1 Supplementary Notes: Equilibrium abundances of the nuclides

The calculations of Burbidge, Burbidge, Fowler & Hoyle [119] on the iron peak equilibrium abundances, referred to in the previous section, are reproduced in Figure 14.2.2. They have been compared to the new solar abundances from the University of Michigan.

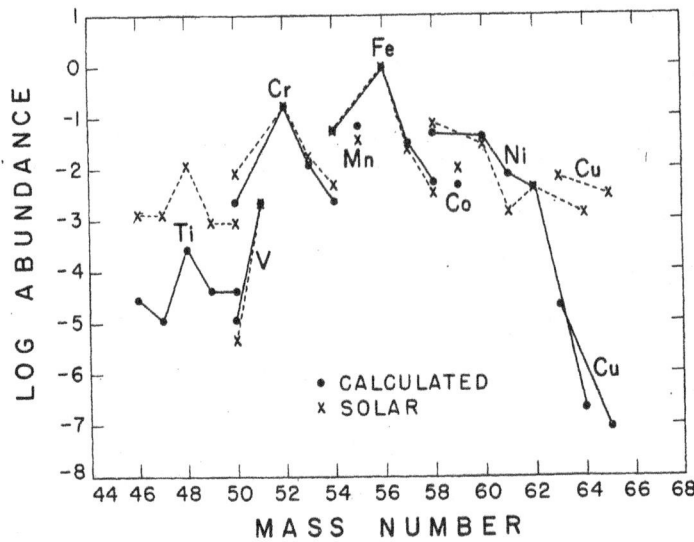

Figure 14.2.2: Equilibrium abundances in the iron peak region, calculated for $T = 3.8 \times 10^9$ °K and $n_n/n_p = 1/300$ [119]. Corrections for freezing reactions were made, and the calculations have been compared with the University of Michigan solar abundance data.

Freezing-in reactions were taken into account. The good agreement was obtained under the condition that the material was relatively proton-rich; this implies that there was not time for full statistical equilibrium to be reached by beta-decay processes.

14.3 The Urca process

Some of the nuclides formed with fairly large equilibrium abundances will be unstable against beta decay. We may consider as an example Mn^{56}, which is unstable against decay into Fe^{56} by 2.8 Mev. In the laboratory, the half-life of Mn^{56} is 2.58 hours, but we will not have this half-life in stars at equilibrium temperatures for two reasons. In the first place, we are very likely to have equilibrium temperatures only at high densities, where the electrons form a degenerate gas. Therefore there is no phase space available into which low energy electrons can be emitted, and the half-life is increased. On the other hand, there will be a considerable equilibrium population of the excited states of

Mn^{56}; some of these can decay to Fe^{56} with a much shorter half-life than the ground state. It seems likely that the latter effect will be more important.

We might think that under equilibrium conditions the beta decay of Mn^{56} would be balanced by electron capture in Fe^{56} produced by bombardment with high energy electrons. However, these reactions are not inverse reactions since a neutrino is emitted when Fe^{56} captures an electron and an antineutrino is emitted when Mn^{56} emits an electron. The neutrinos and antineutrinos escape from the star with negligible nuclear interactions. Hence, the matter and the neutrinos are not in thermodynamic equilibrium, and we do not have detailed balancing in beta decays. However, we must have a general balance between the neutrons and protons in the matter, and this general balance is maintained by positron emission and electron capture in nuclei like Co^{56}. Hoyle [117] estimates that at very high densities (10^{11} g/cm^3), there is a mean time of about 100 seconds required for equilibrium to be set up between the neutrons and protons.

The process by which the star loses energy in the form of neutrino emission has been called the "Urca" process by Gamow & Schoenberg [120]. They computed rates of energy loss by the Urca process, but their calculations are not applicable for two reasons: they assumed detailed balancing between beta emission and electron capture for a given pair of nuclei, and they assumed a nondegenerate electron gas. Nevertheless, the energy loss provided by neutrino emission under equilibrium conditions is very important, in that it can induce a slow collapse of a stellar core, thus bringing about higher central temperatures.

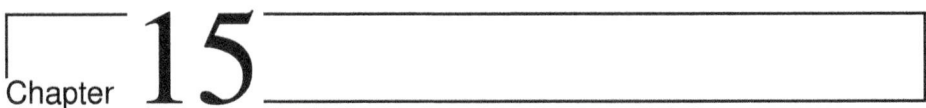

Chapter 15

Supernova Explosions

We have now followed the trend of the nuclear reactions in stellar interiors to very high temperatures and densities. The observational evidence derived from stars with unusual abundances of the elements does not show directly that temperatures above about 1.5×10^8 °K have been reached. However, the equilibrium conditions we have described form the convincing framework for a theory of supernova explosions.

15.1   Observed characteristics of supernovæ

Every few centuries, one of the stars in a galaxy flares up in a gigantic explosion in which much of its mass is thrown off into space with velocities of several thousands of kilometers per second. The light output at maximum can exceed that of the sun by a

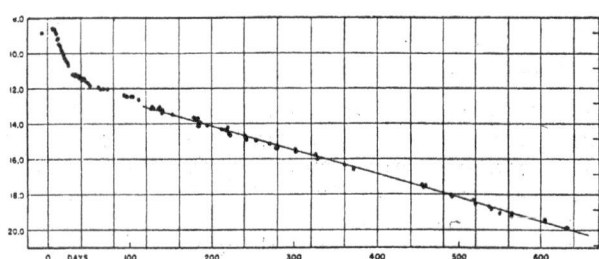

Figure 15.1.1: Photographic light curve of the supernova in IC 4182. The abscissa gives the time after maximum, in days; the ordinate gives the apparent photographic magnitude.

factor of 10^8 or more. Every supernova appears individualistic to some extent, but most of the properties of the explosions can be put into one of two groups.

Type I supernovæ appear to belong to the stars of Population II since only supernovæ of this type appear in elliptical galaxies. In these explosions, the light rises to a maximum in, perhaps, about two weeks after the first rise in the light curve (the rising portion is very poorly observed for obvious reasons). After reaching the maximum, the curve falls rapidly, but after 50 to 100 days, the light curve flattens off, decreasing thereafter at the constant rate of 0.0137 astronomical magnitudes per day. This is an exponential decrease corresponding to a half-life of 55 ± 1 days. The light curve of the supernova in the galaxy IC 4182 is shown in Figure 15.1.1 [121]. This is a particularly significant supernova because it occurred in an unusually small galaxy, and it was brighter than usual. Hence the light curve could be followed very far into the exponential tail; more than 10 half-lives are covered in the figure, and there is no sign of a departure from the exponential decay at the end of this time.

The spectra of Type I supernovæ are like those of no other celestial objects. They consist of broad emission bands (the breadth of the bands being attributed to the Doppler shifts of lines emitted from the ejected gases). These bands have not been identified with known optical transitions. There is no noticeable continuum of emission underlying these bands.

Type II supernovæ appear to belong to Population I stars since they are usually associated with the gas and dust in the arms of spiral galaxies. The rise and early fall of their light curves resemble those of Type I supernovæ, but the rate of fall continues to be very steep, and there is no sign of an exponential tail. Also, the spectra of these objects resemble spectra obtained from nova explosions — the spectra of hot objects with recognizable absorption lines. The total amount of energy released in Type II supernova explosions appears to be somewhat less than in Type I supernova explosions.

15.1.1 Supplementary Notes: Light curves of type I supernovæ

The light curve shown in Figure 15.1.1 has an interesting history. It was first published by Mayall [122]. However, it appears that good modern corrections to the photographic magnitude scale were not established until 1952. The corrections are small at magnitude

16 and large at magnitude 20, being in such a direction as to lower the last part of the tail of the light curve. Thus, it appears that there is no assurance of linearity in the exponential decay beyond about six or seven half-lives of the decay. The suggested direction of departure from linearity is consistent with the Cf^{254} theory if there is a loss of efficiency in the conversion of fission-fragment kinetic energy into visible light. It also removes difficulties expected in connection with the flattening of the light curve owing to the likely presence of Cf^{252}.

15.2   A possible supernova explosion mechanism

Let us consider some of the conditions which must be satisfied before very high central temperatures ($\sim 4 \times 10^9$ °K) can be reached in a star. At such high temperatures, the internal heat content of the matter is enormous: it can be supplied from nuclear reactions (converting hydrogen into iron) only at very high densities. But at very high densities, the electrons form a degenerate gas, which has a very high thermal conductivity.

We have already seen that degenerate stellar cores will tend to be isothermal in the case of stars entering the giant branch in the H-R diagram, and their temperatures will rise appreciably above those of the shell sources supplying the energy generation only at very high luminosities, when the energy release at the center from gravitational contraction becomes large. We have also seen that the degenerate cores must expand when the rise in the central temperature allows helium thermonuclear reactions to start.

The details of stellar evolutionary tracks in the H-R diagram during the helium consumption and subsequent stages are not known. Quite possibly, different stars take quite different paths, depending on such details as the initial angular momentum of the star when formed, as well as upon slight differences in chemical composition. However, we will assume that some of the stars attain high luminosity during or after the consumption of their helium. Such stars then may add mass to an inert core fast enough to cause a rapid central contraction and hence temperatures considerably in excess of 2×10^8 °K. If the star has had this very high luminosity during the helium consumption stage, then the inert core ia probably largely composed of C^{12}.

If the central temperature reaches 6×10^9 °K, carbon thermonuclear reactions will commence. This may cause another core expansion accompanied by internal mixing throughout much of the stellar interior.

It is not clear whether to expect a series of core expansions and contractions as different heavy-ion thermonuclear reactions commence. However, it should be noted that once the central temperature gets much above 10^9 °K, we will have heavy-ion thermonuclear reactions which form nuclei of intermediate weight (neon to calcium). Some of these nuclei are beta-unstable, as may be seen in Table 12.1. At the same time, photonuclear reactions take place which remove alpha-particles from oxygen and neon, as well as from the products of the heavy-ion thermonuclear reactions. Some carbon will be formed from these alpha-particles. The situation will have to be examined in detail to determine the rate of formation of beta-unstable nuclear species as well as the formation of nuclides with small beta stability whose low-lying excited states can capture or emit electrons.

Accompanying these beta transformations will be neutrino and antineutrino emissions. These particles carry energy directly away from the stellar interior. It is very important to determine the rates of neutrino emission from heavy-ion thermonuclear products as functions of temperature and density. Neutrino emission is a powerful cooling mechanism which can induce a slow collapse of the stellar interior, despite the presence of energy-yielding thermonuclear reactions.

If a neutrino-induced collapse of the stellar interior sets in, then it appears that a supernova explosion will inevitably be produced. Accompanying the collapse will be an increased central temperature, which will increase the rate of neutrino emission. Very little additional energy is obtained as the products of the heavy-ion thermonuclear reactions are converted into iron peak nuclei at the higher temperatures where statistical equilibrium sets in.

Hoyle [117] has pointed out that the collapse of the stellar core induced by this Urca process will be a slow process, taking many millions of years. The collapse becomes considerably faster towards its later stages in which the central temperature approaches 5×10^9 °K. At this point the central conditions reach the vicinity of the boundary line in Figure 14.2.1. When the central temperature is increased beyond about 5×10^9 °K, the equilibrium composition of the material must change from favoring Fe^{56} to favoring

He^4.

Now, Fe^{56} is the nucleus with the greatest binding energy per nucleon. The net result of all thermonuclear reactions which produced it was a release of energy, a major fraction of which has been lost from the star by radiation and neutrino emission. When the central temperature crosses the boundary line in Figure 14.2.1, the Fe^{56} must be broken up into alpha-particles, a process which may descriptively be called the photodisintegration of the iron peak. This process requires a large source of energy before it can take place. The only available source of energy is the gravitational potential energy of the core. Hence the transformation from Fe^{56} to He^4 must be accompanied by a very rapid collapse of the stellar core, a collapse which is limited only by the rate at which beta transformations can preserve equilibrium between neutrons and protons. Hoyle [117] calculates that the period of the collapse will be about 100 seconds. A more refined estimate is very desirable.

Let us now consider what happens in the stellar envelope when the core collapses. The outer layers of the envelope may contain a mixture of hydrogen, helium, carbon, nitrogen, oxygen, neon, and the products of heavy-ion thermonuclear reactions. When the underlying support is removed from these layers, they will also fall toward the center of the star. The gravitational potential energy thus released will suddenly heat some of these outer layers to temperatures of the order of 10^8 °K or more. Hydrogen and helium thermonuclear reactions then take place with enormous rapidity, thus raising the temperature still further and increasing the nuclear reaction rates. In other words, a thermonuclear explosion takes place which will blow off these outer layers with velocities of thousands of kilometers per second. This is the visible result of the supernova explosion.

We will proceed to consider the sort of thermonuclear reactions taking place in the supernova envelope. These depend upon the chemical composition of the reacting layers.

Chapter 16

Thermonuclear Reactions on a Fast Time Scale

At the very high temperatures which are reached in thermonuclear explosions in supernova envelopes, there is far less dependence of the reaction rates on the atomic number than is the case at lower temperatures. Therefore it is necessary to consider a very much wider range of hydrogen and helium thermonuclear reactions than we have done so far. Because this topic is a very new one, very few of the necessary calculations have been done.

We shall consider fast time scale reactions in an order which may correspond to a succession of envelope layers lying progressively farther from the collapsing core of the supernova. First we consider regions where hydrogen is absent, then deficient, and finally very abundant.

16.1 Heavy-ion reactions on a fast time scale

The first case to be considered is that in which the supernova envelope layer contains no hydrogen and only moderate amounts of helium at most. We will assume that the temperature of this layer is raised to the range 1 to 2×10^9 °K during the implosion. In this temperature range, the helium will rapidly react to form carbon and heavier nuclei. Carbon thermonuclear reactions will be very rapid. The energy released in these reactions may be sufficient to raise the temperature to about 3×10^9 °K, thus allowing equilibrium conditions to be set up to a good approximation. In this way the nuclei of the iron abundance peak can be formed in this innermost, exploding layer. This is probably the origin of the iron abundance peak which is ejected into space.

16.2 Helium reactions on a fast time scale

In the case where hydrogen is absent but helium is very abundant, and the temperature is raised above about 300×10^6 °K, we will have the helium rapidly consumed. We have seen previously that the higher temperatures favor the $Be^8(\alpha, \gamma)C^{12}$ reaction. Therefore carbon is likely to be the main product of the fast time scale reactions.

Certain other competing reactions may be of interest. It may be that in some cases hydrogen had been exhausted in the imploding layer at temperatures above 100×10^6 °K. We have seen from Figure 11.2.4 that the carbon-cycle nuclei will then be left mostly in the form of N^{13}, which will subsequently decay to C^{13}. If this has happened, then the C^{13} will rapidly interact with alpha-particles during the supernova explosion, thus producing a large number of neutrons. We will consider the fate of these neutrons in Section 16.4.

16.3 Neutron production on a fast time scale

Next we consider the case in which hydrogen is deficient in the imploding layer, having an abundance comparable with the sum of the abundances of carbon, nitrogen, oxygen, neon, and the products of heavy-ion thermonuclear reactions. The hydrogen is then likely to be exhausted at an early stage in the thermonuclear explosion. It is only necessary that the imploding layers be raised to a temperature of about 100×10^6 °K to initiate the explosion. We have previously suggested that this temperature may sometimes be reached without an explosion taking place, but it was always considered that the high temperature of hydrogen exhaustion would be reached in a very small part of the star, as compared to the major fraction of the star which is involved in supernova explosions.

At very high temperatures, protons can interact with some beta-stable nuclei to give exothermic (p, α) reactions. The products of the reactions and other beta-stable nuclei

146

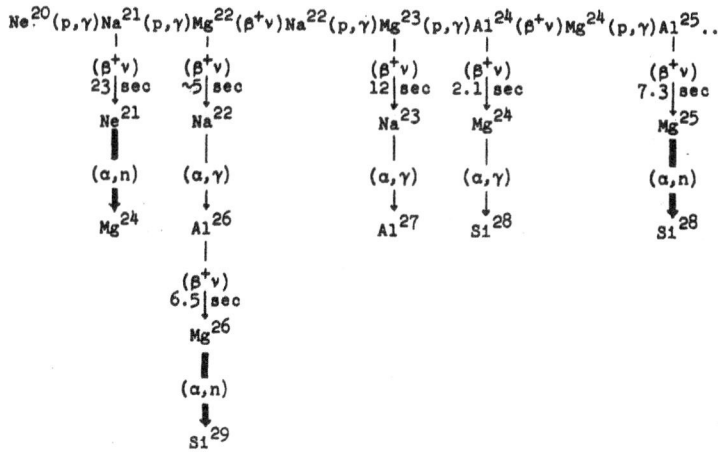

Neutron Production in Supernova Explosions

Figure 16.3.1: A chain of reactions which may possibly occur in supernova explosions. As long as hydrogen is present, proton additions will maintain the horizontal chain of reactions at the top of the figure. After hydrogen is exhausted, the products of proton capture will beta decay and interact with alpha-particles along the vertical lines. Thick arrows show neutron production.

can only capture protons. Very few of the neutron-deficient proton-capture products undergo exothermic (p, α) reactions.

Thus, starting with any light nucleus, proton capture may be followed by the emission of an alpha-particle, but thereafter, in general, only (p, γ) reactions will occur. Very few of these can take place before the proton binding energy drops to such a low value that photoproton reactions prevent further proton additions. The products must then wait until positron emission occurs before further proton captures can take place.

One possible proton capture chain is shown in Figure 16.3.1. Here the initial nucleus is assumed to be Ne^{20}. This nucleus can capture a proton, provided the temperature does not exceed about 600×10^6 °K; otherwise, photoproton reactions will prevent the formation of Na^{21}. The next step is $Na^{21}(p, \gamma)Mg^{22}$; this reaction is probably considerably faster than $Ne^{20}(p, \gamma)Na^{21}$. Hence only a small amount of Na^{21} will be present in the envelope when hydrogen is exhausted. The writer estimates that Al^{23} has zero proton binding energy; hence Mg^{22} must beta decay before the capture chain can continue.

In this way the capture chain shown in the horizontal line at the top of Figure 16.3.1 was constructed. This is not the only possible capture chain; in particular, at higher temperatures the $Al^{24}(p, \gamma)Si^{25}$ reaction will successfully compete with the beta decay of Al^{24}.

After hydrogen is exhausted, some of the neutron-deficient capture products will be left in the exploding layer. Most of these have beta-decay half-lives of a few seconds, so that major portions of them will decay during the period of the supernova explosion. Meanwhile, reactions with alpha-particles are taking place at a slower rate than had been the case for reactions with protons. It may be seen from Figure 16.3.1 that some of the reactions with the beta-decay products produce neutrons.

We shall be interested in the results of neutron capture in heavy elements, to be discussed in the next section. However, in considering the mechanisms of production of neutrons on a fast time scale, it is useful to consider the absorption of the neutrons by (n, p) reactions in light nuclei. We have already seen that the $N^{14}(n, p)C^{14}$ reaction prevents the synthesis of heavy elements by neutron capture on a slow time scale if there is too much N^{14} present. N^{14} is a case which is stable against electron capture by less than the difference between the masses of the neutron and the proton. In all cases where electron capture or positron emission is energetically possible, (n, p) reactions are exothermic. Many such nuclei are produced by proton capture in supernova explosions. In general, the neutron absorption cross sections of these nuclei will be very large, and their abundances will also be very large compared to the abundance of the iron peak.

Most neutrons are produced by exothermic (α, n) reactions from nuclei with mass numbers in the $4A + 1$ series. It has been argued above that the Na^{21} produced by the reactions shown in Figure 16.3.1 would be left with a much smaller abundance than Mg^{22}. The latter nucleus will have a large (n, p) cross section and so will its beta-decay daughter Na^{22}. It is possible that the formation of Si^{25} will provide enough neutrons to allow capture in the nuclei of the iron peak after destroying the nuclei with large (n, p) cross sections. In this case two beta decays must take place before Mg^{25} can be formed and neutrons can be produced. It will be necessary to examine the relative rates of all the reactions considered here before any definite answers can be given to these questions.

The writer believes that the greatest production of neutrons on a fast time scale will occur under conditions in which the hydrogen abundance in the imploding layer is considerably less than the abundance of C^{12}. We have seen that C^{12} is likely to be the main end product of helium thermonuclear reactions at the very high energy production rates which may have occurred in pre-supernovæ. We will assume that this C^{12} has not reacted with the hydrogen mixed with it prior to the implosion.

In this case, when the thermonuclear explosion starts, the hydrogen will be consumed almost entirely in reactions with C^{12}, very little of it reacting with other nuclei. The energy released in these reactions will be assumed to raise the temperature of the material into the range 300 to 600×10^6 °K. If the temperature exceeds about 600×10^6 °K, then it is likely that the N^{13} formed by proton capture in C^{12} will be very rapidly consumed by the reactions $N^{13}(\alpha, p)O^{16}$ and $N^{13}(\gamma, p)C^{12}$.

The beta-decay half-life of N^{13} is ten minutes. Hence about one percent of the N^{13} will decay to C^{13} in the first ten seconds after hydrogen has been exhausted. The C^{13} produced will rapidly interact with helium to yield neutrons. Some of the neutrons will be captured by heavy elements, but it is likely that most of them will be absorbed by N^{13}, giving the reaction $N^{13}(n, p)C^{13}$. In this way the neutrons are regenerated at the expense of the N^{13} which has been formed by proton capture. However, the protons produced in the (n, p) reactions react with the remaining C^{12} to replace the N^{13}. These interaction cycles can therefore probably maintain a large flux of neutrons in the exploding layer for a period of the order of 100 seconds.

16.3.1 Supplementary Notes: Neutron production on a fast time scale

In the previous section, a cycle of reactions was outlined through which it was suggested that neutron capture on a fast time scale took place. This reaction cycle is shown schematically in Figure 16.3.2 if for the moment we neglect the last row of the diagram.

Some further calculations have been carried out to test the plausibility of this postulate. There is no doubt that the neutron production would take place under the postulated set of conditions; the principal question is whether heavy-element synthesis can occur in competition with neutron absorption by the $N^{13}(n, p)C^{13}$ reaction. Average capture cross sections for a sample of the nuclides lying on the fast capture path shown later in Figures 16.4.1 and 16.4.2 have been calculated for an average neutron energy

NEUTRON PRODUCTION IN SUPERNOVAS

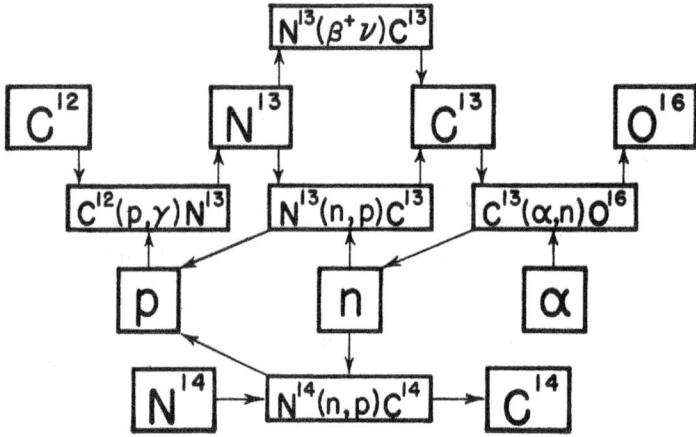

Figure 16.3.2: The postulated cycle of reactions responsible for neutron production on a fast time scale.

of 55 kev. These cross sections lie in the vicinity of one millibarn, a few higher, many considerably lower.

The cross section for the $N^{13}(n, p)C^{13}$ reaction can be inferred from the cross section for the inverse $C^{13}(p, n)N^{13}$ reaction [123]. It is 1100 millibarns at 33 kev neutron energy in the center of mass system.

The synthesis of large quantities of heavy elements requires that initial iron nuclei must capture 100 or more neutrons per nucleus. Hence the N^{13} abundance in the reacting region would have to be considerably in excess of 100 times the abundances of iron plus other medium-weight elements which can be important. Since there is a less than one in 10^3 branching ratio for neutron capture in heavy nuclei, the generalized C^{12}-N^{13}-C^{13} cycle must turn over more than 1000 times for every neutron used for heavy-element synthesis. Hence there must be at least 10^5 times as many C^{12} nuclei as heavy elements and probably closer to at least 10^6 times as many. In the Suess-Urey, abundance table there are only 10^4 C^{12} nuclei per iron plus calcium nucleus if it is assumed that the hydrogen and helium are all converted to C^{12}. Thus the suggested reaction mechanism is inadequate to explain the synthesis of large amounts of Cf^{254}. Reaction mechanisms such as that shown in Figure 16.3.1 are, of course, even more inadequate since the exothermic (n, p) reactions will also have cross sections comparable to 1000 millibarns. Possible reaction cycles similar to that of Figure 16.3.2 but involving a nucleus like Ne^{20} in place of C^{12} will also prove inadequate.

The situation can be rescued if it is assumed that the large amount of C^{12} reacts with the small amount of hydrogen before the supernova explosion takes place. Then, as we saw for the case of neutron capture on a slow time scale, the hydrogen is mostly effective in converting a small amount of C^{12} to C^{13} with only a small amount of the C^{13} being further processed to N^{14}. When the explosion starts, neutrons are produced from C^{13}. Now the only competition with heavy-element synthesis is the $N^{14}(n, p)C^{14}$ reaction. This has an s-wave neutron cross section at 55 kev of only 1.2 millibarns. The p-wave cross section is probably not much larger. The protons released by this reaction initiate the rest of the cycle discussed above, but since the amount of N^{14} is small and its cross section is small it appears that this more generalized set of reactions can account for the synthesis of large amounts of Cf^{254}.

Although the suggestion of pp. 148-149 cannot account for the synthesis of large amounts of Cf^{254}, it may be responsible for the conversion of many nuclei formed by neutron capture on the slow time scale to nuclei which are products of the fast time scale. Only a small amount of neutron capture in the heavy elements is required in this case. This possibility complicates the problem of the analysis of the reaction mechanisms responsible for the abundances of the heavy nuclei.

Hoyle (private communication) has suggested another mechanism of neutron production on a fast time scale which may be of importance in nucleogenesis. Prior to the collapse of a supernova core, the primary constituent of the core is Fe^{56}. If this is suddenly converted to He^4, there will be four neutrons left over. During the course of the explosion, some of the material may be ejected into cooler regions through the dynamics of the implosion. The neutrons can then be captured by the trace of material which has not disintegrated to He^4 or by the small amount of C^{12} which forms from recombination of the He^4. These possibilities have not been quantitatively explored.

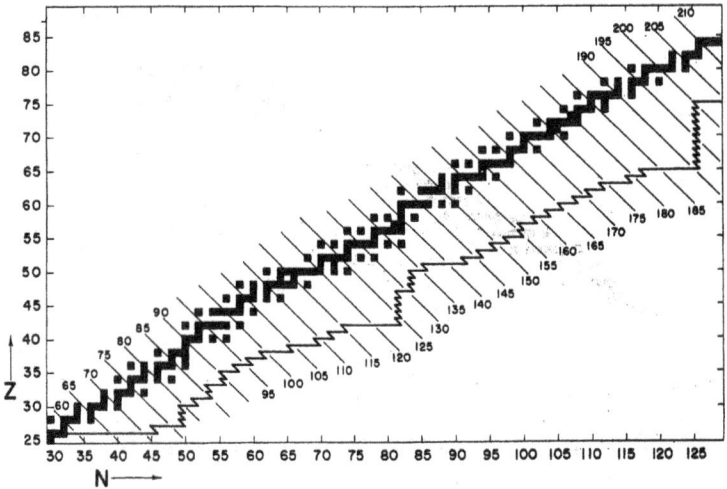

Figure 16.4.1: A nuclide chart showing the path of neutron capture on a fast time scale constructed on the assumption that the photodisintegration will prevent neutron capture when the neutron binding energy falls below 2 Mev. The blackened squares show beta-stable nuclides. The diagonal lines show mass numbers. The capture has been assumed to start in Fe^{56}.

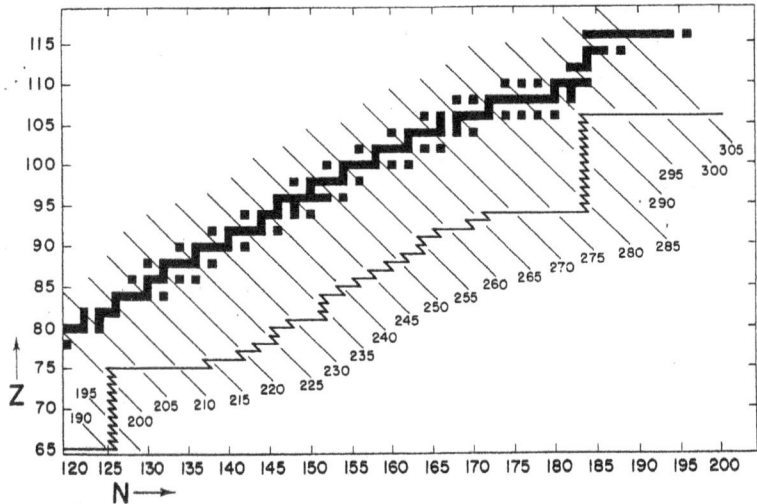

Figure 16.4.2: A continuation of Figure 16.4.1 for very heavy nuclei.

16.4 Neutron capture on a fast time scale

We have seen previously that heavy nuclides have much larger (n, γ) cross sections than light nuclides. These cross sections are probably of the same order of magnitude as, but somewhat less than, the $N^{13}(n, p)C^{13}$ cross section. Hence neutrons will be captured by heavy nuclei at a faster rate than by light nuclei during a supernova explosion.

We will consider two extreme sets of conditions which may control the course of the neutron capture in heavy nuclei. We will call these the low-flux and the high-flux conditions.

In the low-flux case, a heavy nucleus will capture a series of neutrons until it has become very neutron-rich, and its beta-decay half-life has become short enough to be comparable with the average time between neutron captures. Beta decay then competes with neutron capture. The beta-decay product captures more neutrons until a position of short beta-decay half-lives is reached again. In this process, the equilibrium abundances of the neutron-capture products are all more or less comparable, but with even-even nuclei having greater equilibrium abundances than odd-mass-number nuclei on account of their smaller neutron-capture cross sections. There will also be increased abundances of nuclides with closed neutron shells.

In the high-flux case, successive neutron additions to a heavy nucleus may occur with a mean interval of some milliseconds. This time is much shorter than any possible beta-decay half-life in the heavy region. Therefore, the neutron capture continues until the neutron binding energy falls so low that photoneutron emission rates become comparable with the neutron-capture rates. When this happens, the heavy nuclide must wait until a beta decay occurs. Generally speaking, the beta-decay half-lives of these neutron-rich nuclei will be in the range 0.1 to 1 seconds, corresponding to decay energies of 10 to 15 Mev. Further neutron addition up to the photodisintegration limit will follow the beta decay. Thus, in the high-flux case, the nuclei with large equilibrium abundances will be those at the photodisintegration limits; their abundances will be inversely proportional to their beta-decay half-lives. There will, on the average, be only one such nucleus per three mass numbers along the neutron-capture path. The other nuclei on the path will have very much smaller equilibrium abundances.

We shall see later in our discussion of cosmic abundances that there is good evidence for the presence of a high-flux process in the synthesis of the heavy elements. Therefore we shall concentrate our attention on this case.

We have seen that photodisintegration rates are very sensitive to particle binding energies. This fact allows us to construct a neutron-capture path for heavy nuclei in a crude way by allowing us to locate approximately the photodisintegration limits along the path.

Let us assume a temperature of about 600×10^6 °K. This is the highest temperature at which one can envisage having neutron production by the $C^{13}(\alpha, n)O^{16}$ reaction, since N^{13} will rapidly be destroyed at this temperature. The resulting protons, obtained from both the $N^{13}(\alpha, p)O^{16}$ and $N^{13}(\gamma, p)C^{12}$ reactions, will form substances with large (n, p) cross sections, which may rapidly absorb the remaining neutrons, thus quenching the high neutron flux very quickly. We will be interested, therefore, in constructing the equilibrium capture path in which the mean time between neutron captures and the mean photodisintegration time are both of the order of one second. This may roughly represent the conditions in which the heavy-element abundances will be "frozen" at the termination of the neutron capture.

Since the nuclei at the photodisintegration limit have small neutron binding energies, their nuclear level spacings are large and their radiation widths are small. Hence such nuclei would have average neutron-capture cross sections at 11 kev much smaller than the values shown in Figure 10.2.3. Let us assume a value of ten millibarns for this cross section. Then it may be seen from equation (13.0.6) that $\lambda \approx 1$ s^{-1} for a neutron binding energy of 2 Mev. Hence, a very crude neutron-capture path corresponding to freezing-in conditions in the heavy nuclei region can be constructed by assuming that neutron capture will take place if the neutron binding energy is greater than 2 Mev, but not if it is less.

The writer has constructed such a neutron-capture path using neutron binding energies computed from his revised semi-empirical atomic mass formula [109]. This formula contains an empirically determined shell correction energy term, and hence the nuclear shell dependences of the neutron binding energies are reproduced in an approximately realistic fashion. The capture path is shown in Figures 16.4.1 and 16.4.2.

In Figure 16.4.1 neutron capture is assumed to start in Fe^{56}. Many neutrons are quickly added, and the nuclei reach the region of 60 neutrons. At this point there is a big drop in neutron binding energies; hence, there follows a series of beta decays interspersed with single neutron captures which maintain the neutron number at 50. This gives a series of nuclei with adjacent mass numbers and with large equilibrium abundances. The half-lives of the beta decays will be short and there will be little or no odd-even mass number variation in their values. The nuclei now approach the valley of beta stability, and their beta-decay half-lives increase, thus giving the nuclei larger equilibrium abundances. Eventually, the binding energy of the 51st neutron exceeds 2 Mev, and rapid neutron addition commences again, to be periodically interrupted when the photodisintegration limit is reached. Similar large equilibrium abundance regions are reached when the neutron number reaches 82, 126, and, by assumption, 184. The capture path for the heaviest region is shown in Figure 16.4.2.

Large equilibrium abundances for nuclides with 184 neutrons are shown in Figure 16.4.2. When beta decay produces the nucleus with atomic number 106 and mass number 289, neutron addition is found from the calculations to be very rapid again. It must be emphasized that this picture is entirely dependent on the assumption of a major closed shell at 184 neutrons. However, assuming the closed shell to be correctly located, it is doubtful that the capture path will reach element 106. The capture path has, by that time, approached the valley of beta stability fairly closely and the nuclei on it have values of Z^2/A comparable to that of Cf^{252}. Z^2/A is a parameter which is correlated with the fissionability of nuclei. Thus the neutron-capture path will probably be terminated by fission somewhere in the vicinity of atomic number 103 and mass number 287.

It is not clear exactly how the fission will occur. The fission widths are larger for the higher excited states of the nuclei, and, hence, spontaneous fission is unlikely to be the mechanism terminating the neutron-capture chain. Excited states are produced following neutron capture and also following beta decay. It is therefore likely that fission produced by neutrons or following beta decay terminates the chain. The latter process might be called "delayed fission."

Swiatecki [124] has found a correlation between the asymmetry of mass splitting in fission and Z^2/A. Hence we may expect the nucleus 103 to split into fragments with about the same mass ratio (1.34) as in the fission of Cf^{252}. This would give fission fragment peaks centered about mass numbers 123 and 164. The fission fragments would capture more neutrons and follow the capture path already discussed.

16.4.1 Supplementary Notes: Neutron capture on a fast time scale

The capture path followed by nuclei for the case of a very high neutron flux has also been discussed by Burbidge, Burbidge, Fowler & Hoyle [1], and it is interesting to note that the conclusions reached by these authors are essentially the same as those obtained in this report. However, there are some differences in their conclusions, and it is instructive to discuss these differences.

Ghiorso [125] has prepared a plot of spontaneous fission half-life against neutron number of the nucleus in which he notes that there is a decrease in the half-lives of the even isotopes of any element when the neutron number exceeds 152. Burbidge *et al.* make long extrapolations of these trends to obtain a spontaneous fission half-life of about 10^{-3} seconds for mass number 260, essentially independent of the charge number. The writer does not believe that these extrapolations are valid. Since nuclear fissionability generally increases as Z^2/A increases, he believes that the spontaneous fission half-lives of the even-even isotopes of a given element will reach a minimum value at a few neutrons above 152 and that the half-lives will then increase as further neutrons are added. He believes that the neutron-capture path lies in a region of sufficiently small Z^2/A that fission will not occur until the capture path approaches the line of beta stability at the hypothetical closed shell of 184 neutrons.

These different interpretations affect the interpretation of the small peak in the abundance distribution near mass number 164. Since the odd-even effect vanishes at this hump, its formation must be attributed to neutron capture on a fast time scale. In this report the feature has been attributed to a fission product peak formed from decay of the heavy nuclei in the abundance peak near $A = 287$, following termination of neutron capture. This interpretation is not available to Burbidge *et al.* because only nuclei in the mass number range 254 to 260 will fission after the neutrons are cut off, and the abundances of these nuclei are low. Instead, Burbidge *et al.* attribute the hump to the "stabilization" of the nucleus in the region where it changes from spherical to deformed as the rare-earth region is entered. By this, it is meant that neutron binding energies are larger than usual but no explanation is given of how such stabilization will increase the equilibrium abundances of the nuclei on the fast capture path. In fact, the writer would expect precisely the opposite effect. The increased neutron binding energy should bring the nuclei farther from the valley of beta stability at the photodisintegration limit. Their beta-decay energies are then larger; their beta-decay half-lives are shorter, and their equilibriuim abundances should be less than those of their neighbors. In order to get an abundance peak, the capture path should approach the valley of beta stability, and it may be seen from Figure 16.4.1 that this does not occur in the region of $A = 164$.

This same point has a bearing on the expected abundances of uranium and thorium in nature. Burbidge *et al.* have argued that "stabilization" effects beyond the closed shell of 125 neutrons will produce a similar abundance peak in the region of thorium, uranium, and the nuclei of slightly larger mass number which decay to thorium and uranium isotopes. The writer would not expect a peak of this sort to be formed.

The writer also differs with Burbidge *et al.* in estimating the half-lives for beta decay for nuclei on the fast capture path. Burbidge *et al.* idealize the problem by assuming a large transition probability for the beta decay to occur to one of the excited states of the daughter nucleus. In fact, beta decay can occur to all the excited states of the daughter nucleus which are energetically accessible and which have acceptable values of spin and parity. The transition matrix elements are likely to be much less, on the average, than that assumed by Burbidge *et al.*, but when one sums over all possible transitions, the half-lives come out considerably shorter, particularly in the region of lighter nuclei. In this way, the writer feels that neutron capture on a fast time scale will synthesize heavy nuclei much faster than thought by Burbidge *et al.*

16.5   Californium 254 and Type I supernovæ

Let us now consider what happens after the neutron supply is shut off in an exploding layer of the type we have been considering. The neutron-rich nuclei approach the valley of beta stability by a series of beta decays, at first with short half-lives of the order of a second but with progressively longer half-lives as the first position of beta stability is approached. Certain beta-stable nuclides are made in this way, which were not made by neutron capture on a slow time scale.

The behavior of very heavy nuclei is of particular interest. Mass numbers 210 to 231 decay into a region of the valley of beta stability which is unstable to alpha-particle

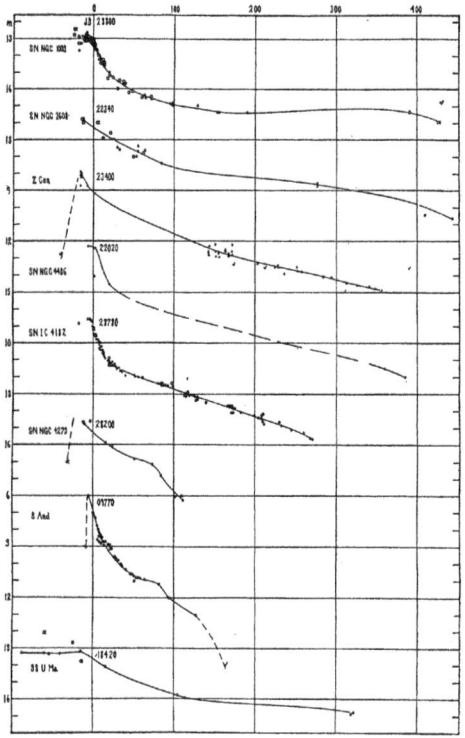

Figure 16.5.1: A compilation of supernova light curves by D. Hoffleit.

emission, the longest half-lives involved being one to two weeks for mass numbers up to 225 and many years for the higher mass numbers. The alpha-decay products eventually become Pb^{206}, Pb^{207}, Pb^{208} and Bi^{209}.

In the mass number region 250 to 260, the beta-stable nuclides begin to decay by spontaneous fission with progressively larger branching ratios. Hence all the very heavy nuclei which have been formed will fission. The large abundance peak expected near mass number 287 will give fission fragments, if Swiatecki's suggestion is correct, near mass numbers 123 and 164.

In the intermediate range of mass number, 232 to 256, the half-lives for alpha decay of the beta-stable nuclides are rather long. The products formed after 10^9 years are Th^{232}, Bi^{209}, U^{238} and U^{235}. It will not be possible to predict the relative abundances of these products until the neutron-capture path can be constructed with certainty and in detail, and the details of the freezing-in process can be determined. This requirement is too exacting to be met by nuclear theory in its present state of development.

Burbidge, Hoyle, Burbidge, Christy & Fowler [126] have pointed out that the product with mass number 254, Cf^{254}, has very interesting and significant properties. It decays by spontaneous fission with a half-life of 55 days. There appears to be no other nucleus formed which both decays predominantly by spontaneous fission and has a half-life nearly as long as Cf^{254}. Since Cf^{254} releases about 200 Mev in its decay, it can be expected to release far more energy than any other product about three months after the explosion.

Following the supernova explosion, the layers of gas expand into space. They quite soon become transparent to radiation, and the thermal energy released in the explosion is radiated away into space.

The remaining gases are then very cool and the atoms probably combine to form molecules. Fission fragments and other decay particles must be slowed down following their ejection, and, in so doing, they will excite the molecules and atoms with which they make close collisions. It is tempting to ascribe the peculiar emission bands of Type I supernovæ to this particle excitation process. At the same time, the exponential decay of the light curve with a half-life of 55 days can be attributed to the spontaneous fission of Cf^{254}, which releases much more energy than any other process.

However, it is necessary to make a critical examination of the question of the possible presence of other activities in the Type I supernova light curve. We can see from an examination of Figure 15.1.1 that the straight line passes slightly under the points until a time of about 100 days from the explosion has passed. The slight additional light in this early period may possibly come from the alpha-particle emissions by substances with half-lives of the order of 1 to 2 weeks, followed by more rapid alpha-particle emissions from the products so formed.

It is a cause of considerable concern that the light curve of Figure 15.1.1 is still a straight exponential after 600 days. Consider what would happen if Cf^{252} and Cf^{254} were formed in equal abundance by the fast time scale process. The latter will have decayed by 11 half-lives at 600 days; its initial abundance is decreased by a factor 2000. Cf^{252} has a half-life of two years and, hence, has fallen in abundance by a factor two. However,

Cf^{252} also has a two percent branching ratio for spontaneous fission. Hence, after 600 days, the energy yield from Cf^{252} should exceed that from Cf^{254} by a factor of 20, and the light curve should have flattened out. However, there is no sign of flattening, and hence Cf^{252} would have to be much less than five percent as abundant as Cf^{254} immediately after the explosion, if the proposed explanation of the light curve is correct.

This problem is possibly solved, in principle, by the observation given previously that, in the region of the neutron-capture path which includes mass numbers 252 and 254, only about one mass number in three has a large equilibrium abundance, provided we have a high flux of neutrons. It remains to be shown in detail that there is a photodisintegration limit at $A = 254$, but none at $A = 252$. It has so happened that this is the case in Figure 16.4.2, but this is in a region of extrapolation of the writer's mass formula, and hence it does not constitute proof that $A = 254$ will be a photodisintegration limit. It is also necessary that the neutron supply be cut off very suddenly so that the abundances will not be locally smeared out during the freezing-in process.

Burbidge *et al.* [126] have found that the amount of Cf^{254} in the light curve of the supernova in IC 4182 corresponds to about two percent of the iron in one solar mass. Now, presumably, this supernova was a star of Population II with much less iron than the sun, a mass not much greater than the sun, and the neutron capture probably did not involve more than ten percent of the mass. Also, only a small fraction of the neutron-capture products would be left with $A = 254$. This would indicate that the neutron capture transformed more than just the iron peak in this supernova; further heavy nuclei were probably formed from neutron capture in the products of heavy-ion thermonuclear reactions and by fission.

We have seen that the conditions required for a large Cf^{254} to Cf^{252} ratio appear to be quite critical. The high flux of neutrons must be terminated very abruptly at a rather specific temperature. It is perhaps questionable whether these conditions will always be well satisfied in Type I supernovæ. A compilation of supernova light curves is given in Figure 16.5.1 [127]. It may be seen that the light curves for the supernovæ in NGC 4486 and Z Centauri resemble that of the supernova in IC 4182 out to about 400 days, although there is no guarantee that they did not subsequently flatten out quite quickly. The light curves of the supernovæ in NGC 1003 and NGC 2608 appear to be flatter than that of the supernova in IC 4182, and other radioactivities may have been prominent in these light curves.

16.5.1 Supplementary Notes: Half-life of californium 254

We have seen in Section 16.5 that the end portion of Baade's light curve for the supernova in IC 4182 must be considered to be very doubtful. Therefore, the half-life of the exponential tail is 55 days but with probably a larger error than the ± 1 day assigned by Baade.

There have now been two determinations of the half-life of Cf^{254} in the laboratory. Using debris from the November 1952, thermonuclear test, Huizenga & Diamond [128] have found the Cf^{254} half-life to be 56.2 ± 0.7 days. Using Cf^{254} made in a reactor, Thompson & Ghiorso (S. G. Thompson, private communication) find the half-life to be 61 days with a similarly small error. These two determinations are thus in disagreement. Dr. Thompson suggests that the sample from the thermonuclear test may have contained some Cf^{256}. If the Ghiorso diagram of spontaneous fission half-lives is already turning up at Cf^{256}, this contaminant might alter the half-life measured by Huizenga & Diamond. Of course, we should realize then that the supernova may also be expected to produce Cf^{256}, which may account for its light curve half-life being close to the measurement of Huizenga & Diamond. Burbidge *et al.* have suggested that the half-life may be altered by the decay of Fe^{59} with a half-life of 45 days. Large amounts of Fe^{59} may be produced in regions where light positron emitters absorb most of the neutrons which have been produced.

We must also consider the possibility, first suggested by Borst [129], that the decay curve is associated with the 53 day half-life of Be^7. We have seen that considerable amounts of Be^7 can be formed at higher temperatures by the $He^3(\alpha, \gamma)Be^7$ reaction. However, in order to account for the energy in the exponential tail in the visible light region, an amount of Be^7 comparable to a solar mass would have to be formed. The hypothesis is therefore unacceptable.

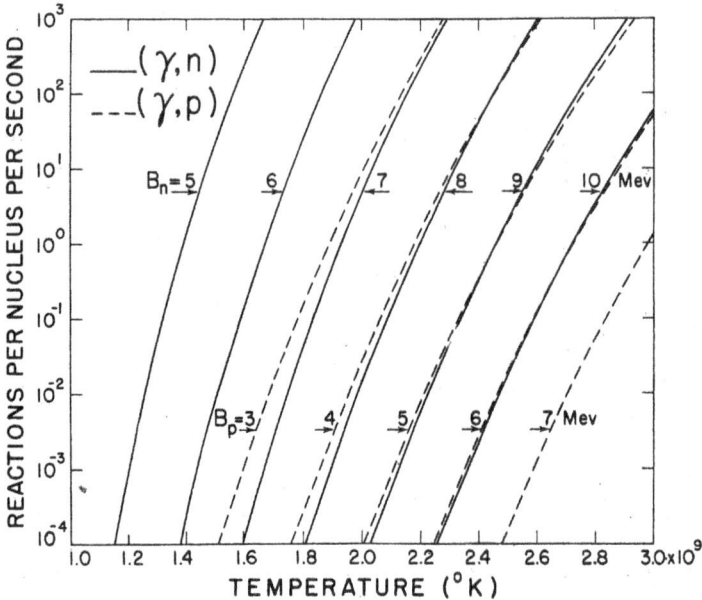

Figure 16.6.1: Photoneutron and photoproton disintegration rates plotted for various binding energies for typical nuclei in the rare-earth region.

16.6 Reactions in regions of large hydrogen abundance

If the hydrogen abundance is large in the supernova exploding layers, then protons will be rapidly added to light nuclei until the proton photodisintegration limit is reached. The nuclei must then undergo a beta decay before further proton addition can take place. This situation is similar to that in which neutrons produced in a high flux are captured on a fast time scale. None of the products formed undergo exothermic (α, n) reactions, and, hence, neutron production is negligible in the present case.

When the temperature is in the range 500 to 1000×10^6 °K, certain photodisintegration limits prevent the addition of protons to some light beta-stable nuclei (*e.g.* C^{12}, O^{16}, Ne^{20}, ...). This prevents such nuclei from being built up to larger mass numbers by proton addition. Nevertheless there appears to be a range of temperature in the vicinity of 500×10^6 °K in which the nuclei in the range from neon to calcium will be synthesized by proton capture on a fast time scale.

At temperatures in the vicinity of 10^9 °K, proton capture is prevented at several

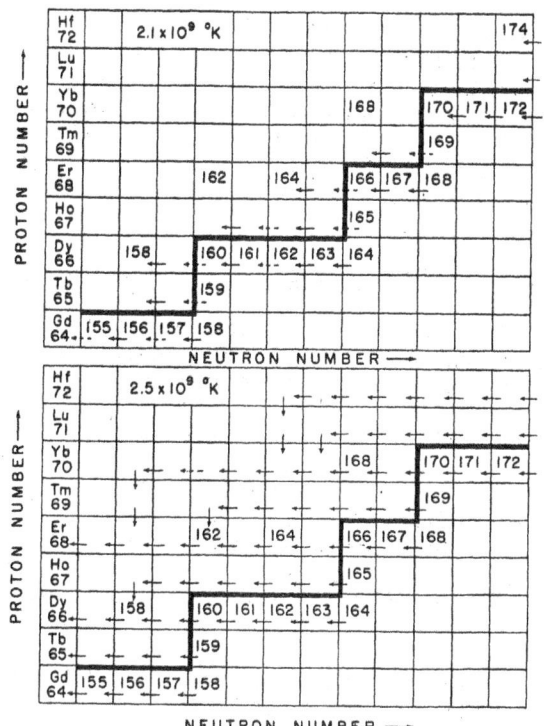

Figure 16.6.2: The course of photodisintegration reactions at temperatures of 2.1×10^9 °K and 2.5×10^9 °K. Solid arrows are reactions with mean reaction times less than one second; dashed arrows correspond to mean reaction times between one and ten seconds. The heavy black line shows the position of the main capture path for neutron capture on a slow time scale. Mass numbers are inserted showing nuclei which are beta-stable. The left-hand column shows the element symbol and atomic number.

photodisintegration limits in nuclei lighter than Ca^{40}. However, proton capture can take place in heavier nuclei at these higher temperatures. In this way, certain beta-stable nuclei can be formed which lie on the neutron-deficient side of the valley of beta stability and which cannot be formed by neutron capture at all.

At much higher temperatures, in the vicinity of 2×10^9 °K, photonuclear reactions start to dominate the scene. Photodisintegration reaction rates corresponding to several neutron and proton binding energies are shown in Figure 16.6.1. The neutron emission rates were taken from Table 13.1; the proton emission rates were calculated for a typical rare-earth nucleus using equation (13.0.3). It may be seen from this figure that there

163

is an approximate equality between the photoneutron and photoproton emission rates when the proton binding energy is 4 Mev less than the neutron binding energy (in the rare-earth region).

The course of the photodisintegration reactions in the rare-earth region is shown in Figure 16.6.2 which shows photodisintegration paths on a section of a nuclide chart that includes the rare-earth region. It has been assumed that large abundances of heavy nuclides have been formed in the star by neutron capture on a slow time scale and have been mixed into the exploding layer. At a temperature of 2.1×10^9 °K, a few neutrons would be removed from these nuclei during the course of the explosion. However, at a temperature of 2.5×10^9 °K, neutron emission is very rapid and continues until proton emission becomes more probable. One or two proton emissions are then followed by further neutron emissions. In this way, the heavy nuclei are rapidly broken down into the vicinity of the iron peak. Certain beta-stable nuclei on the neutron-deficient side of the valley of beta stability are formed which cannot be formed by neutron capture.

It seems reasonable to postulate that supernovæ of Type II are those in which no high neutron fluxes have been produced, probably owing to large hydrogen excess conditions throughout their envelopes. Such supernovæ contain for the most part only very short-lived radioactivities and hence would not have exponential tails in their light curves.

Analysis of Nuclide Abundances

In the course of the preceding discussion, we have seen that all nuclei existing in nature can be formed in stellar interiors under conditions which may lead to ejection into space, with the exception of deuterium, lithium, beryllium, and boron nuclei. These exceptions will be discussed later. However, it is not sufficient just to show that nuclei can be formed in stars and ejected to the interstellar medium: before one can have any confidence in a theory of nucleogenesis which incorporates these mechanisms, it is necessary to show that the abundances of the nuclides existing in nature are consistent with the abundances which would be produced by the postulated mechanisms. We shall now see that this appears to be the case.

17.1   Abundances of very light nuclei

We consider first the abundances of the nuclei in the range carbon to neon. These nuclei are transformed by hydrogen thermonuclear reactions on a slow time scale.

The carbon-cycle nuclei are not in carbon-cycle equilibrium proportions on the earth, as has been mentioned previously. C^{13} and N^{14} are in approximate carbon-cycle proportions relative to each other, but C^{12} and N^{15} are then relatively overabundant. However, we have seen that C^{12} is independently produced by helium thermonuclear reactions and N^{15} is independently produced (probably in Type II supernovæ) by hydrogen reactions on a fast time scale.

There is a very large abundance of O^{16}, which is one of the main products of helium reactions. The abundance of O^{17} is very small, possibly corresponding to the fact that it is destroyed by hydrogen reactions. The natural abundance may have been mostly produced by hydrogen reactions on a fast time scale. O^{18} has an abundance small compared to O^{16} but large compared to O^{17}. It may have been produced by the reaction $N^{14}(\alpha, \gamma)F^{18}(\beta^+\nu)O^{18}$ in helium reactions on both fast and slow time scales. A similar situation exists with respect to F^{19}, which can be made by the $O^{15}(\alpha, \gamma)Ne^{19}(\beta^+\nu)F^{19}$ reaction. In this case, only a fast time scale reaction would be satisfactory because only under such conditions is an appreciable abundance of mass number 15 present in the reacting medium.

The relative abundances of the neon isotopes are similar to those of the oxygen isotopes, and it is likely that they would be produced by a similar series of reactions: $O^{16}(\alpha, \gamma)Ne^{20}$, $Ne^{20}(p, \gamma)Na^{21}(\beta^+\nu)Ne^{21}$, and $Ne^{18}(\alpha, \gamma)Mg^{22}(\beta^+\nu)Na^{22}(\beta^+\nu)Ne^{22}$.

17.2 Abundances of intermediate nuclei

We have seen that the nuclei in the range from neon to calcium or titanium can be produced by several mechanisms:

(a) Neutron capture on both slow and fast tine scales.

(b) Heavy-ion thermonuclear reactions on both slow and fast time scales.

(c) Hydrogen thermonuclear reactions on a fast time scale.

(d) Helium thermonuclear reactions on a fast time scale.

Fowler, Burbidge & Burbidge [63] suggested that neutron capture on a slow time scale might account quite successfully for the abundances of these nuclei. The reason for this is that the products of neutron-capture cross sections and abundances are the same within an order of magnitude or so, and they have a slight monotonic decrease with increasing mass number, as would be expected for a neutron capture process without a very close approach to equilibrium. However, this suggestion cannot be valid unless there is a strict monotonic decrease in the products, and this is not true. For example, the product for neutron capture in Al^{27} is an order of magnitude higher than the corresponding products in Mg^{25} and Mg^{26}. Hence neutron capture on a slow time scale cannot give a major contribution to the abundances of intermediate nuclei, although they undoubtedly make a significant one.

Not enough is known about the details of the various processes mentioned above to make a good analysis of their effects on the abundances of intermediate nuclei. The small abundance of S^{36} suggests that neutron capture on a fast time scale has not been too important, but this may also be just the result of high-flux conditions in which mass number 36 may have a very small equilibrium abundance. The abundances of nuclei containing an integral number of alpha-particles are generally larger than those of other nuclei, but this could have resulted not only from successive alpha-particle captures but also from the effects of proton reactions on a fast time scale in which fast photoproton reactions prevent proton addition to these nuclei. It is evident that much work remains to be done to clarify the situation in the intermediate range of mass number.

17.3 The iron abundance peak

We have discussed in some detail the formation of an iron abundance peak under equilibrium conditions, and we have seen that Hoyle, Fowler, Burbidge, & Burbidge [118] have obtained a remarkably good agreement between the calculated equilibrium abundances and those observed in nature.

17.4 Abundances of heavy odd-mass-number nuclei

When we pass beyond the iron peak into the region of heavy nuclei, we encounter products of nuclear reactions which can be made in only a few ways. These nuclei have abundances which are susceptible to an analysis which can determine the relative importance of the various production mechanisms.

In Figure 17.4.1 are plotted the abundances of heavy nuclides with odd mass number according to Suess & Urey [102]. There are several distinctive features in this plot. The sharp abundance spikes near mass numbers 90 and 140 correspond to the large abundances of nuclei formed by neutron capture on a slow time scale which have closed shells

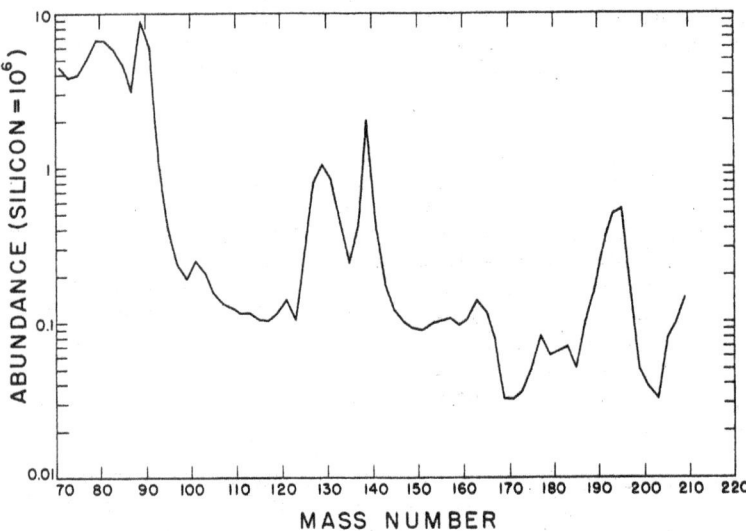

Figure 17.4.1: The abundances of heavy nuclides with odd mass numbers according to Suess & Urey [102].

of 50 and 82 neutrons. A similar abundance spike is to be expected corresponding to 126 neutrons; it would be present if the solar abundance of lead [60] had been incorporated in this plot as it was in Figure 9.4.1.

There are also broad abundance peaks centered about mass numbers 80, 130, 195. C. D. Coryell [130] has suggested that these peaks correspond to the effects of closed neutron shells of 50, 82, and 126 neutrons in neutron capture on a fast time scale. We can see confirmation of this view in Figure 16.4.1, in which large equilibrium abundances are produced in just these mass number positions. A detailed comparison between Figure 16.4.1 and Figure 17.4.1 shows that in the region of mass number 80 the abundances seem to correspond to a limiting neutron binding energy slightly more than 2 Mev, and, in the region of mass number 195, the abundances seem to correspond to a limiting neutron binding energy slightly less than 2 Mev. This is the sort of behavior to be expected since, for a given neutron flux, higher temperatures raise the photodisintegration limit energies and slow down the rate of neutron capture by heavy nuclei. The abundances of lighter nuclei may therefore be expected to reflect greater temperatures and higher photodisintegration limiting energies.

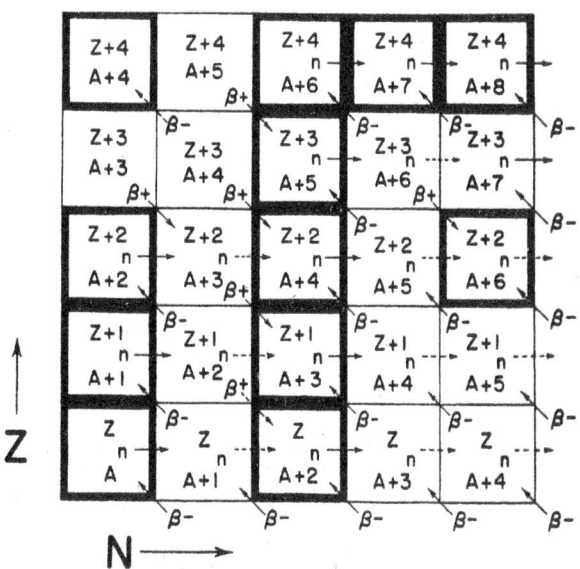

Figure 17.5.1: A typical section of a nuclide chart showing beta-stable nuclei outlined with heavy black borders.

The next most prominent feature in Figure 17.4.1 is the subsidiary peak centered about $A = 163$. It is tempting to identify this peak with the peak predicted in Section 16.5 for the fission of the nuclei, which would be left over after neutron capture ceases in an abundance peak at about $A = 287$. We indicated previously that these fissions may give fission fragments with masses centered about $A = 123$ and $A = 164$. The region at $A = 123$ is obscured in Figure 17.4.1 by the rise of the broad peak at $A = 130$.

The smaller abundances present at other positions in Figure 17.4.1 presumably are formed by neutron capture on slow and fast time scales, but we must examine the abundances of nuclear isobars in order to separate the effects of different mechanisms.

17.5   Abundances of heavy isobars

In the region of the heavy nuclei, there is often more than one beta-stable nuclide with the same mass number, particularly for even mass numbers. Figure 17.5.1 shows a typical section of a nuclide chart in which we will be able to define three classes of these isobars.

Let us first consider the pair of isobars $(Z, A + 2)$ and $(Z + 2, A + 2)$. The latter

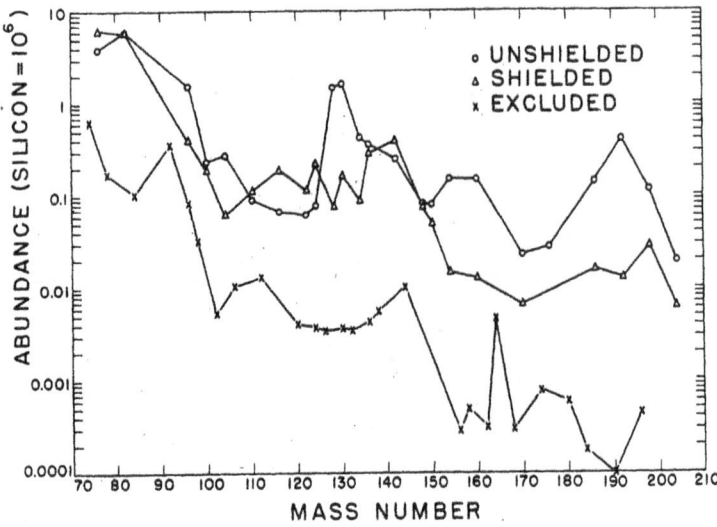

Figure 17.5.2: The abundances of the three classes of isobars according to Suess & Urey [102].

will be formed by neutron capture on a slow time scale if the beta-decay half-life of the nucleus $(Z, A + 1)$ is short compared to the mean time between neutron captures in $(Z, A+1)$. This beta-decay half-life will be called the "shielding half-life." If it is small, then $(Z + 2, A + 2)$ will be a product of the slow time scale, and $(Z, A + 2)$ will be a product of the fast time scale. Hence, $(Z, A + 2)$ will be called an "unshielded isobar," and $(Z + 2, A + 2)$ will be called a "shielded isobar." Also in this figure, $(Z + 2, A + 6)$ is an unshielded isobar, and $(Z + 4, A + 6)$ is a shielded isobar.

The nuclide $(Z + 4, A + 4)$ cannot be formed by neutron capture at all. It will be called an "excluded isobar." We have seen that nuclides of this sort can be formed by proton capture on a fast time scale and by photonuclear reactions.

The abundances of these three classes of isobars according to Suess & Urey are shown in Figure 17.5.2. In this figure only isobars have been plotted in which, if formed by neutron capture, the shielding half-life is less than 1 year. If very hot stars pass through a neutron production stage very quickly, then some products of neutron capture on a slow time scale may have been included in the unshielded isobar abundances, particularly in the heavy region $A > 170$, where the shielding half-lives are usually some weeks.

The unshielded isobars have abundance peaks at $A = 130$ and $A = 192$, as is to be expected from Figure 17.5.1. The shielded isobars do not have peaks at these places, but there is a peak at $A = 140$, as is also to be expected. The general level of abundances of the shielded isobars decreases by a factor of about five in going above $A = 140$. We have seen that this behavior is also to be expected for neutron capture on a slow time scale.

The excluded isobars are presumably formed by proton capture and photonuclear reactions starting with the products of neutron capture on a slow time scale. Hence, one would expect that the general trend of the abundances of the excluded isobars should parallel that of the shielded isobars. This is seen to be the case in Figure 17.5.2. Let us examine the excluded isobars with $A > 150$. Four of these have relatively much larger abundances than the other six. These four nuclides can be formed by photonuclear reactions on both fast and slow time scales. The other six can only be formed by photonuclear reactions on a fast time scale. This suggests that proton capture has not been important in this region. The clear abundance separation into two time scale classes is very gratifying.

For $A < 150$, no such irregular abundance structure is observed; the abundances are smooth functions of mass number. Hence, these excluded isobars have presumably been formed for the most part by proton capture on a fast time scale.

17.6   The odd-even abundance effect

Further support for the mechanism identifications made above can be obtained by observing the fluctuations in the abundances of adjacent mass numbers, shown in Figure 9.4.1. Neutron capture on a slow time scale produces a very jagged appearance in the resulting abundances, as may be seen in Figures 11.1.1 to 11.1.8. Neutron capture on a fast time scale has two associated effects which produce a much smoother variation of abundances with mass number:

(a) In high-flux conditions, the abundances of nuclei with closed shells of neutrons are inversely proportional to the beta-decay half-lives which should have very little odd-even effect.

(b) After neutron capture ceases, the fission products of the heavy peak, presumably

171

at $A = 287$, should form two peaks with very little odd-even variation. These peaks were predicted at $A = 123$ and $A = 164$.

It may be seen that there is a large odd-even variation in the peak at $A = 80$. Hence much of this peak must be due to products of the slow time scale, as well as the fast. The variation in the peak at $A = 130$ is much reduced; in the peak at $A = 195$, variation is almost nonexistent. This behavior is consistent with expectations. Finally, we may note that there is a great reduction in the odd-even variation in the regions of $A = 123$ and $A = 164$. This is consistent with the presence of fission products in these regions.

It is possible that the products of neutron capture on a fast time scale may exhibit a reduced odd-even effect owing to smearing of abundances in the freezing-in process and the confounding of abundances from different sources with different conditions. However, it appears that the fast-time-scale abundances may be relatively unimportant, except in the mass-number regions discussed here.

Chapter 18

Nuclear Reactions in Stellar Surfaces

In our considerations of nuclear reactions in stellar interiors, we have found production mechanisms by which the elements heavier than boron may have been synthesized in our galaxy. We have not accounted for the abundances of deuterium, lithium, beryllium, and boron. Nor have we accounted for the abundance anomalies observed in certain types of stars. We will now consider nuclear reactions which may take place in stellar surfaces, and we shall see that these remaining puzzles in nuclear astrophysics may be related and may find their explanation in stellar surface reactions. These reactions are believed to be associated with strong magnetic activity in stellar surfaces.

Stellar surface reactions were first discussed by Fowler, Burbidge & Burbidge [93], who associated these reactions with the acceleration of charged particles by the betatron effect in growing starspots. The following discussion borrows heavily from the paper of Fowler and the Burbidges, but somewhat different mechanisms of acceleration are proposed, and the actual nuclear reactions regarded as of major importance are somewhat different. We must first consider the nature of magnetic activity in the sun and stars.

18.1   The magnetic field of the sun

The sun has a weak dipole field with a polar field strength of the order of 1 gauss. In the equatorial regions, the dipole nature of the general field is much distorted by local irregularities and strong local fields. Field strengths of several hundred gauss are

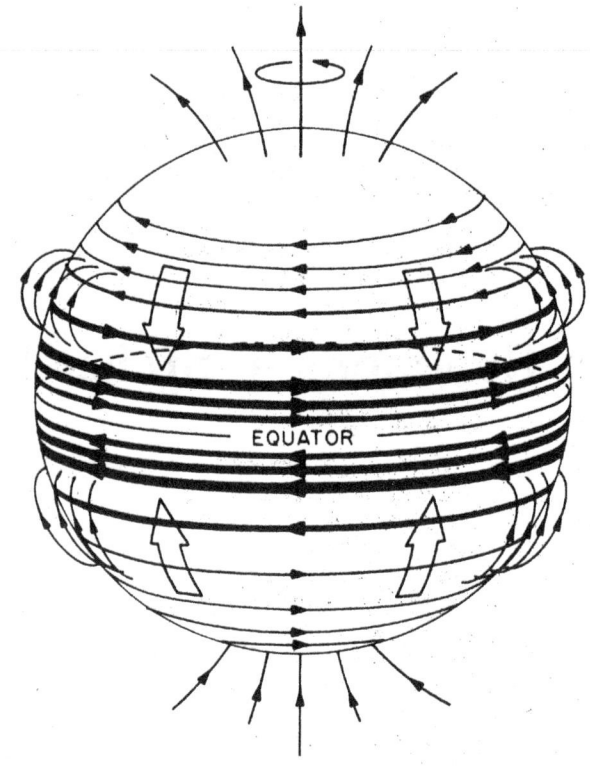

Figure 18.1.1: The nature of the sun's magnetic field according to the dynamo theory [131].

commonly observed in sunspots. Intermediate field strengths are observed in unipolar and bipolar magnetic regions.

A modern theory of the sun's magnetic field has been summarized by Parker [131]. It is believed that the field is generated by hydromagnetic dynamo action in the sun. The basis of the dynamo is a strong, internal toroidal field, illustrated in Figure 18.1.1. This toroidal field interacts with the motion of the matter in the outer convection zone of the sun to generate a weak poloidal field. The lines of force in the poloidal field are, in turn, drawn out by material motions to strengthen the toroidal field. The sun appears to go through an oscillation in which a new toroidal field is generated in the polar regions every 11 years in the opposite sense to the toroidal field previously generated. The alternating bands of oppositely oriented toroidal fields move slowly toward the sun's equator, where

174

they vanish.

It appears that sunspots are produced when a bundle of magnetic lines of force in the toroidal field floats up to the surface of the sun. This produces two sunspots (or spot groups) which differ in magnetic polarity and are joined by flux lines which arch through the sun's atmosphere between the spots. In the course of time, the spots subdivide and rotate slowly, and the magnetic lines of force become twisted and tangled. Often when the spots are growing or declining most rapidly, flares are produced, which evidently represent points at which large electrical breakdowns occur, possibly caused by the twisting and tangling of the magnetic lines of force.

18.2   Magnetic variable stars

About 13 percent of all stars in the spectral range B8 to F0 show spectral peculiarities which appear to be caused by abundance anomalies. These stars appear to be just evolving away from the main sequence, so that it is unlikely that any helium reactions have started in their cores. It appears very probable that all these stars have strong surface magnetic fields. The field strengths, averaged over a hemisphere, are usually several thousand gauss. The stars are often variable, with periods of a few days and light curves of small amplitude. During the period, the magnetic field is usually observed to be variable and often changes sign. There are associated changes in the strengths of the absorption lines of different elements. These properties have recently been reviewed by Deutsch [132].

The abundance peculiarities of these peculiar A stars will be discussed in more detail in Section 18.5, but here it is worth stating that oxygen and calcium appear to be underabundant, the metallic group comprising the iron peak is slightly overabundant as would be expected for young Population I stars, the elements near mass number 90 are overabundant by factors of about 25, and the rare earths and lead are overabundant by factors of several hundred or several thousand (except for barium which appears to have about normal abundance). The elements of intermediate mass number generally have nearly normal abundances, except for silicon which is overabundant by a factor of about ten. It may be seen that these abundance anomalies are not characteristic of neutron capture on a slow time scale.

The most intense of the peculiar lines are those of the elements Si, Cr, Mn, and Eu. The line strengths of Si, Cr, and Mn vary together with those of Fe during the magnetic cycle (although there are some individual characteristics for different elements). The strontium lines vary in the same phase. However, the Eu lines and those of the other rare earths vary 180° out of phase with the above lines.

Most of the previous statements refer specifically to the star α^2 Canum Venaticorum, analysed in detail by Burbidge & Burbidge [133]. So far as is known, they appear to be generally applicable to peculiar A stars, although not all such stars show strengthening of the lines of all the elements mentioned.

There is, as yet, no satisfactory model for these stars. Among the suggestions which have been made, are that the principal dipole axis of the magnetic field is inclined at an angle to the axis of rotation. This oblique rotator model is capable of explaining the reversal of magnetic field strength associated with the rotation of the stars, but it fails to explain many other details of the spectrum variability. Hence a physical variation in the magnetic field is also required. Since, in highly ionized cosmic objects, the lines of magnetic flux are strongly glued to the material in which they exist, then magnetic variability must be associated with physical motions of the gases at the surfaces of the stars. It is not clear whether a successful model can be constructed using such physical motions alone or whether oblique rotation is also required. In any case, it is clear that the abundance peculiarities are not evenly distributed over the surface of these stars. The rare-earth group of elements seems to occur in patches, and some of the other elements with anomalous abundances may also be unevenly distributed.

18.3 Nuclear reactions in flares

It was indicated in Section 18.1 that solar flares are probably the visible effects of large electrical discharges in regions between sunspots where the magnetic lines of force have become twisted and tangled at times when the sunspots are growing or shrinking most rapidly. It seems not unlikely that similar discharges will be much more frequent and much more violent in the atmospheres of peculiar A stars where the magnetic field is enormously greater, and physical motions are likely to distort it much more rapidly than in the sun. It may be that extensive regions on peculiar A stars may be at times in a state of nearly continuous flaring.

Since there are rapid changes in magnetic field strengths at the times of flares, it is very likely that some charged particles will be accelerated by the betatron effect in regions of fairly low density (10^8 or 10^9 particles per cm^3) where ionic collisions are not too frequent. It is also likely that there are strong hydromagnetic shock waves moving through the flare regions. Charged particles will be reflected from such shock fronts, gaining energy if the shock front was moving towards the particle and losing energy if the shock front was overtaken from the rear. For either type of acceleration, a power-law energy spectrum of charged particles will be produced, the number of particles of energy E varying approximately as $E^{-2.5}$ [93]. At high energies (~ 1 Bev), the spectrum must fall off much more rapidly owing to the short time during which the flare is in existence (see in this connection Meyer, Parker & Simpson [134]).

One of the characteristics of this power-law spectrum is that heavier nuclei undergoing acceleration will tend to be stripped of electrons early in the acceleration, and hence, in general, these nuclei will be nearly as efficiently accelerated (in terms of energy imparted per nucleon) as the singly charged particles. As a result, these heavy particles will penetrate the Coulomb barriers of other heavy nuclei nearly as readily as the singly charged particles will.

Protons of a few Mev will be elastically scattered from other protons and from helium nuclei and will be partly elastically and inelastically scattered from heavier nuclei. A great deal of neutron production will occur by (p, n) reactions, smaller amounts by (p, 2n), (p, pn), (p, 3n) and similar reactions induced by the high energy tail of the power-law spectrum. These neutrons, for the most part, probably diffuse from the flare before being captured; most will be captured by hydrogen, forming deuterium. If these deuterons are later accelerated then they may, for the most part, capture protons in collisions, forming He^3. These reactions are of the type usually called spallation reactions. Lithium, beryllium, and boron nuclei will be spallation products from proton collisions with nuclei in the carbon, nitrogen, and oxygen group.

Deuterons and alpha-particles accelerated to low energies of a few Mev may add

177

particles by (d, p), (d, n), (α, n), (α, 2n), and (α, 3n) reactions. At higher energies, more particles will be emitted, and hence such nuclei then also will initiate spallation reactions.

Also of interest may be collisions between heavy particles. Thus oxygen collisions with oxygen would give a highly excited form of S^{32} which would then evaporate a few nucleons. The ultimate products would be silicon and phosphorus nuclei. Collisions between iron nuclei would produce nuclei with mass numbers in the vicinity of 90 to 100.

It has been suggested that most of the tritium found in the earth's atmosphere may have been produced by spallation reactions in the sun, followed by the ejection of clouds of gas from the solar surface [135].

18.3.1 Supplementary Notes: Deuterium in a solar flare

L. Goldberg (to be published*) has measured the profile of the Hα line emitted by a solar flare and has found an asymmetrical hump in it, which can be explained if ten percent of the hydrogen in the flare region is deuterium. This appears to indicate that nuclear activity can be very large in sunspot regions.

18.4 Regions of high kinetic temperature

If we now consider regions of gas with higher densities ($\sim 10^{15}$ particles per cm^3), then we find that a power-law spectrum of particles will not be produced owing to the much smaller collision mean free paths of the particles. Instead, the energy imparted by the acceleration will raise the kinetic temperature of the plasma region in which it occurs.

Let us consider nuclear reactions in a plasma region in which the kinetic temperature has been raised to more than 10^{10} °K, or a few Mev in energy units. High temperature plasmas of this general nature are the current goal of controlled fusion research [137]. Such plasma regions must be confined by very high magnetic fields, of the sort found in peculiar A stars, but no speculations are given here about the mechanisms for production of such regions.

Collisions between ions in high temperature plasmas maintain a Maxwell distribution of energies corresponding to the kinetic temperatures involved. When the electrons

---

*Editor's Note: Now published [136].

collide with these ions, they radiate bremsstrahlung. This has a powerful, cooling effect which maintains the temperature of the electrons below that of the ions. The plasma can be maintained at a high kinetic temperature only if the mechanism of generation continues to pump energy into the region or if the ions gain enough energy from exothermic nuclear reactions. The latter possibility seems somewhat questionable owing to the additional cooling effect of the endothermic (p, n) reactions.

Because of the Maxwell distribution of the ionic energies, there will be very few reactive collisions between heavier ions with multiple charges. Hence we need consider only reactions induced by singly charged ions. We will assume that the gas has an initial appreciable abundance of deuterium produced by neutron capture in hydrogen.

Protons of a few Mev bombarding heavier elements may be captured, or at higher energies, they may produce (p, n) reactions. Either type of reaction will produce nuclei on the neutron-deficient side of the valley of beta stability. Collisions with deuterium will tend to add a particle by (d, p) and (d, n) reactions, but the former type of reaction will predominate in heavier nuclei. These reactions will, therefore, tend to produce nuclei on the neutron rich side of the valley of beta stability. The net result of these two types of reactions is to capture a series of protons and neutrons and to build up heavy nuclei from light ones.

The details of the abundances of heavier nuclei produced by these processes will depend critically upon the nuclear reaction energies involved and upon the Coulomb barrier heights of the bombarded nuclei. Reactions with lighter nuclei will occur much faster that with heavier nuclei; hence heavy nuclei, perhaps in the rare-earth region, will be the principal products remaining when the reactions cease. It is possible that there may be "feedback" loops involving (p, α) and (d, α) reactions setting in at the closed shell of 82 neutrons which will maintain a high abundance of heavy nuclei immediately beyond barium, but this question has not yet been investigated in detail.

The very light nuclei of the carbon, nitrogen, and oxygen group are likely to be broken down by (p, α) reactions in a high temperature plasma. Among the more prominent products of these reactions are Li^7 and B^{11}.

Table 18.1: Atomic abundance ratios for two peculiar A stars.

Element	$\frac{\alpha^2 \text{ CVn}}{\gamma \text{ Gem}}$	$\frac{\text{HR5597}}{\alpha^2 \text{ CVn}}$
Mg	0.4	3.5
Al	1.1	2.0
Si	10	2.5
Ca	0.02	2.6
Sc	0.7	—
Ti	2.6	1.0
V	1.3	2.3
Cr	5.2	1.8
Mn	16	1.0
Fe	2.9	1.5
Ni	3.0	0.8
Sr	14	1.1
Element	$\frac{\alpha^2 \text{ CVn}}{\text{sun}}$	$\frac{\text{HR5597}}{\alpha^2 \text{ CVn}}$
Y	20	—
Zr	30	1.3
Ba	0.9	—
La	1020	0.2
Ce	400	0.6
Pr	1070	0.6
Nd	250	0.6
Sm	410	0.6
Eu	1910	0.5
Gd	810	0.4
Dy	760	0.6

18.5 Abundances in peculiar A stars

The most complete analyses of peculiar A stars are due to Burbidge & Burbidge [119; 133]. They determined the abundances of lighter elements in the peculiar A stars α^2 Canum Venaticorum and HR 5597 relative to the normal star γ Geminorum as a standard and the heavier elements relative to the sun as a standard. Their results are given in Table 18.1. If all the abundance features of this table are correct, then the following conclusions seem reasonable.

The small abundance of calcium suggests that the atmosphere of these peculiar A stars has been extensively processed through regions in which spallation reactions take place. Most of the calcium abundance is in the isotope Ca^{40}; spallation reactions in which a few nucleons are removed on the average will therefore reduce the abundance of calcium by a large factor.

These same spallation reactions would also produce large quantities of Mn^{55} from Fe^{56}, either directly or through production of Fe^{55}. The large abundance of iron would not be lost immediately because of the formation of Fe^{54} from Fe^{56}, which is the isotope of major abundance.

The overabundance of silicon may result from extensive collisions of oxygen with oxygen. This element is the third most abundant in stars, next to hydrogen and helium. It is known that oxygen is deficient in α^2 Canum Venaticorum.

The overabundances of Sr, Y, and Zr may result from collisions between nuclei of the iron abundance peak with each other.

All the abundance anomalies just discussed seem to be associated with each other in the sense that the lines of these elements vary essentially in phase with each other during the magnetic cycle of α^2 Canum Venaticorum. However, europium and the rare earths vary out of phase with the aforementioned elements. Since the rare earths seem to be those most localized in patches in these stars, it is therefore tempting to postulate that the rare earths have been produced by reactions in high temperature plasmas. This is also the only explanation which shows promise of accounting for the normal abundance of barium.

18.6   Lithium stars

Certain giant carbon stars show very strong lines of lithium. Spitzer [138] has estimated that lithium is overabundant by a factor of forty in one of them.

It is not known whether these giant stars have strong magnetic fields. However, strong magnetic fields on the surfaces of red giant stars are not uncommon, and it may be that these carbon stars also have such fields. In this case it is possible that the excess lithium has been produced by spallation reactions.

18.7   Red dwarf flare stars

We have seen that deuterium, lithium, beryllium, and boron can be produced by spalla-tion reactions and, to some extent, also by fusion reactions in hot plasmas. However, it is quantitatively unlikely that the peculiar A stars can have produced enough of these elements to account for their cosmic abundances, simply because there are relatively few peculiar A stars in space.

However, it has been observed that nearly all the very faint dwarf stars of class M have frequent large flares of a few minutes duration. The structures of these stars are not well known theoretically as yet, but it appears certain that they have very deep convection zones which may extend inwards to the center from the surface. They may have strong magnetic fields. Since they are the most numerous type of star in space, it may be that the large cosmic abundances of D, Li, Be, and B have been formed by spallation reactions in the flares in these stars, followed by ejection into space before mixing into the interior has occurred.

18.8   Supplementary Notes: T Tauri stars

Hunger & Kron [139] have observed the emission of polarized light frcan the T Tauri star NX Monocerotis. This is a member of the class of stars which gives periodic flares and which appears to be still in the gravitational contraction phase of its formation from the inter-stellar medium. The polarized light is characteristic of synchrotron radiation from electrons moving in strong magnetic fields. It seems not improbable that this star is still in the process of ridding itself of its initial strong magnetic field and that many nuclear reactions can take place on the stellar surface during this process.

Presumably, every star may go through a similar stage during its initial contraction phase. If so, then the formation of the cosmic abundances of D, Li, Be, and B would be very much easier to explain as the result of surface magnetic activity. It is also conceivable that the solar system abundances of these elements could be higher than in the inter-stellar medium, if they were formed by the sun in the gases from which the planets were later formed.

Bibliography

[1] Burbidge, E. M., Burbidge, G. R., Fowler, W. A., & Hoyle, F., "Synthesis of the Elements in Stars." *Reviews of Modern Physics* **29** (1957) 547–650.

[2] Cameron, A. G. W., "Nuclear Reactions in Stars and Nucleogenesis." *Publications of the Astronomical Society of the Pacific* **69** (1957) 201–222.

[3] Zwicky, F., "Statistics of Clusters of Galaxies. Distribution of Centers, Angular Dimensions, Structure, Luminosity Function of Member Galaxies," in "Proceedings of the Third Berkeley Symposium on Mathematical Statistics and Probability, Volume 3: Contributions to Astronomy and Physics," ed. J. Neyman, University of California Press, 1956 pp. 113–144.

[4] Terrien, J., "News from the International Bureau of Weights and Measures." *Metrologia* **4** (1968) 41–45.

[5] Russell, H. N. & Moore, C. E., *The Masses of the Stars: With a general catalogue of dynamical parallaxes*, The University of Chicago Press, 1940.

[6] Hoyle, F., "Remarks on the computation of evolutionary tracks," in "Stellar populations: Proceedings of the Conference Sponsored by the Pontifical Academy of Science and the Vatican Observatory, May 20-28, 1957," ed. D. O'Connell, Ricerche astronomiche, North Holland Publishinging Co, 1958 pp. 223–230.

[7] Deutsch, A. J., "The Circumstellar Envelope of Alpha Herculis." *The Astrophysical Journal* **123** (1956) 210–228.

[8] Salpeter, E. E., "The Luminosity Function and Stellar Evolution." *The Astrophysical Journal* **121** (1955) 161–167.

[9] Friedman, A., "Über die Krümmung des Raumes." *Zeitschrift fur Physik* **10** (1922) 377–386.

[10] Bondi, H. & Gold, T., "The Steady-State Theory of the Expanding Universe." *Monthly Notices of the Royal Astronomical Society* **108** (1948) 252–270.

[11] Hoyle, F., "A New Model for the Expanding Universe." *Monthly Notices of the Royal Astronomical Society* **108** (1948) 372–382.

[12] Hoyle, F., "On the Cosmological Problem." *Monthly Notices of the Royal Astronomical Society* **109** (1949) 365–371.

[13] Alpher, R. A. & Herman, R. C., "Theory of the Origin and Relative Abundance Distribution of the Elements." *Reviews of Modern Physics* **22** (1950) 153–212.

[14] Gamow, G., "Expanding Universe and the Origin of Galaxies." *Det Kongelige Danske Videnskabernes Selskab, matematisk-fysiske Meddelelser* **27** (1953) 1–15.

[15] Baum, W. A., "Photoelectric determinations of redshifts beyond 0.2 c." *The Astronomical Journal* **62** (1957) 6–7.

[16] Hoyle, F., "On the Fragmentation of Gas Clouds Into Galaxies and Stars." *The Astrophysical Journal* **118** (1953) 513–528.

[17] Sandage, A., "The Systematics of Color-Magnitude Diagrams and Stellar Evolution." *Publications of the Astronomical Society of the Pacific* **68** (1956) 498–500.

[18] Walker, M. F., "Studies of Extremely Young Clusters. I. NGC 2264." *The Astrophysical Journal Supplement Series* **2** (1956) 365–387.

[19] Walker, M. F., "A study of extremely young clusters of stars." *The Astronomical Journal* **62** (1957) 37–37.

[20] Baum, W. A., "Globular Clusters Observed through a Crystal Ball." *Smithsonian Contributions to Astrophysics* **1** (1956) 165–176.

[21] Greenstein, J. L., "The Spectra and Other Properties of Stars Lying below the Normal Main Sequence," in "Proceedings of the Third Berkeley Symposium on Mathematical Statistics and Probability, Volume 3: Contributions to Astronomy and Physics," ed. J. Neyman, University of California Press, 1956 pp. 11–29.

[22] Sandage, A. *Mémoires of la Société Royale des Sciences de Liège* **14** (1954) 254.

[23] Luyten, W. J., "The Spectra and Luminosities of White Dwarfs." *The Astrophysical Journal* **116** (1952) 283–290.

[24] Chandrasekhar, S., *An Introduction to the Study of Stellar Structure*, University of Chicago Press, 1939.

[25] Chandrasekhar, S., "The structure, the composition, and the source of energy of the stars," in "Astrophysics, a Topical Symposium: Commemorating the Fiftieth Anniversary of the Yerkes Observatory and a half century of progress in Astrophysics," ed. J.A. Hynek, McGraw-Hill Book Company, 1951 pp. 508–681.

[26] Thompson, W. B., "Thermonuclear Reaction Rates." *Proceedings of the Physical Society B* **70** (1957) 1–5.

[27] Wigner, E. P. & Eisenbud, L., "Higher Angular Momenta and Long Range Interaction in Resonance Reactions." *Physical Review* **72** (1947) 29–41.

[28] Yost, F. L., Wheeler, J. A., & Breit, G., "Coulomb Wave Functions in Repulsive Fields." *Physical Review* **49** (1936) 174–189.

[29] Teichmann, T. & Wigner, E. P., "Sum Rules in the Dispersion Theory of Nuclear Reactions." *Physical Review* **87** (1952) 123–135.

[30] Lane, A. M., *Harwell report A.E.R.E.* **T/R 1289** (1954).

[31] Lane, A. M. & Thomas, R. G., "R-Matrix Theory of Nuclear Reactions." *Reviews of Modern Physics* **30** (1958) 257–353.

[32] Blatt, J. M. & Weisskopf, V. F., *Theoretical Nuclear Physics*, John Wiley & Sons Inc, 1952.

[33] Porter, C. E. & Thomas, R. G., "Fluctuations of Nuclear Reaction Widths." *Physical Review* **104** (1956) 483–491.

[34] Salpeter, E. E., "Nuclear Reactions in the Stars. I. Proton-Proton Chain." *Physical Review* **88** (1952) 547–553.

[35] Schatzman, E., "Les Réactions Themonucléaires aux Grandes Densités (Gaz Dégénérés et non Dégénérés)." *Le Journal de Physique et le Radium* **9** (1948) 49.

[36] Schatzman, E., "Sur certaines réactions nucléares d'importance astrophysique. IV. Sensibilité des réactions thermonucléaires." *Annales d'Astrophysique* **16** (1953) 162–178.

[37] Schatzman, E., "On the Effect of Electron Screening on Thermonuclear Energy Generation." *The Astrophysical Journal* **119** (1954) 464–466.

[38] Keller, G., "The Effect of Electron Screening on Thermonuclear Energy Generation." *The Astrophysical Journal* **118** (1953) 142–146.

[39] Salpeter, E. E., "Electrons Screening and Thermonuclear Reactions." *Australian Journal of Physics* **7** (1954) 373–388.

[40] Heller, L., "Theories of Element Synthesis and the Abundance of Deuterium." *The Astrophysical Journal* **126** (1957) 341–356.

[41] Fowler, W. A. *Mémoires of la Société Royale des Sciences de Liège* **14** (1954) 88.

[42] Marion, J. B. & Fowler, W. A., "Nuclear Reactions with the Neon Isotopes in Stars." *The Astrophysical Journal* **125** (1957) 221–232.

[43] Bethe, H. A. & Critchfield, C. L., "The Formation of Deuterons by Proton Combination." *Physical Review* **54** (1938) 248–254.

[44] Frieman, E. & Motz, L., "The Proton-Proton Reaction and Energy Production in the Sun." *Physical Review* **83** (1951) 202.

[45] Feenberg, E. & Trigg, G., "The Interpretation of Comparative Half-Lives in the Fermi Theory of Beta-Decay." *Reviews of Modern Physics* **22** (1950) 399–406.

[46] Holmgren, H. D. & Johnston, R. L., "He$^3(\alpha, \gamma)$Be^7 Reaction." *Bulletin of the American Physical Society* **3** (1958) 26.

[47] Holmgren, H. D. & Johnston, R. L., "H$^3(\alpha, \gamma)$Li^7 and He$^3(\alpha, \gamma)$Be^7 Reactions." *Physical Review* **113** (1959) 1556–1559.

[48] Dunning, K. L., Butler, J. W., & Bondelid, R. O., "Mass and Half-Life of B^8." *Bulletin of the American Physical Society* **1** (1956) 328–329.

[49] Breit, G. & Yost, F. L., "Radiative Capture of Protons by Carbon." *Physical Review* **48** (1935) 203–210.

[50] Thomas, R. G., "An Analysis of the Energy Levels of the Mirror Nuclei, C^{13} and N^{13}." *Physical Review* **88** (1952) 1109–1125.

[51] Wilkinson, D. H., "Special Mechanisms in the Reaction ^{7}Li(p, γ)^{8}Be." *Philosophical Magazine* **45** (1954) 259–276.

[52] Wichers, E., "Report on Atomic Weights for 1956-1957." *Journal of the American Chemical Society* **80** (1958) 4121–4124.

[53] Cameron, A. G. W., "Origin of Anomalous Abundances of the Elements in Giant Stars." *The Astrophysical Journal* **121** (1955) 144–160.

[54] Öpik, E. J., "Stellar Structure, Source of Energy, and Evolution." *Publications of the Tartu Astrofizica Observatory* **30** (1938) 1.

[55] Öpik, E. J., "Secular changes of stellar structure and the ice ages." *Monthly Notices of the Royal Astronomical Society* **110** (1950) 49–68.

[56] Öpik, E. J., "Stellar Models with Variable Compositions II. Sequences of Models with Energy Generation Proportional to the 15th Power of Temperature." *Contributions from the Armagh Observatory* **3** (1951).

[57] Öpik, E. J., "Stellar Models with Variable Compositions II. Sequences of Models with Energy Generation Proportional to the 15th Power of Temperature." *Proceedings of the Royal Irish Academy Section A* **54** (1951) 49.

[58] Bosman-Crespin, D., Fowler, W. A., & Humblet, J., "Sur le Taux de génération d'énergie thermonucléare das les étoiles." *Bulletin de la Société Royale des Sciences de Liège* **9** (1954) 327–339.

[59] Bethe, H. A., "Energy Production in Stars." *Physical Review* **55** (1939) 434–456.

[60] Goldberg, L., Müller, E. A., & Aller, L. H., "The chemical composition of the solar atmosphere." *The Astronomical Journal* **62** (1957) 15–16.

[61] Lamb, W. A. & Hester, R. E., "Transmutation of Nitrogen by Protons from 100 kev to 135 kev." *Physical Review* **108** (1957) 1304–1307.

[62] Warren, J. B., Laurie, K. A., James, D. B., & Erdman, K. L., "Gamma Radiation from the Proton Bombardment of Oxygen." *Canadian Journal of Physics* **32** (1954) 563–570.

[63] Fowler, W. A., Burbidge, G. R., & Burbidge, E. M., "Stellar Evolution and the Synthesis of the Elements." *The Astrophysical Journal* **122** (1955) 271–285.

[64] Wilkinson, D. H., "Radiative transitions in light elements: II." *Philosophical Magazine* **1** (1956) 127–152.

[65] Broström, K. J., Huus, T., & Koch, J., "Gamma-Ray Yield Curves of Separated Neon Isotopes Bombarded with Protons." *Nature* **160** (1947) 498–500.

[66] Newton, T. D., "Shell Effects on the Spacing of Nuclear Levels." *Canadian Journal of Physics* **34** (1956) 804–829.

[67] Pixley, R. E., Hester, R. E., & Lamb, W. A. S., "Radiative Capture of Protons in Ne^{20}." *Bulletin of the American Physical Society* **2** (1957) 377.

[68] Henyey, L. G., "The Evolution of Stars Near the Main Sequence." *Publications of the Astronomical Society of the Pacific* **68** (1956) 503–504.

[69] Greenstein, J. L. & Richardson, R. S., "Lithium and the Internal Circulation of the Sun." *The Astrophysical Journal* **113** (1951) 536–546.

[70] Greenstein, J. L. & Tandberg-Hanssen, E., "The Abundance of Beryllium in the Sun." *The Astrophysical Journal* **119** (1954) 113–119.

[71] Schwarzschild, M., Howard, R., & Härm, R., "Inhomogeneous Stellar Models. V. a. Solar Model with Convective Envelope and Inhomogeneous Interior." *The Astrophysical Journal* **125** (1957) 233–241.

[72] Hoyle, F. & Schwarzschild, M., "On the Evolution of Type II Stars." *The Astrophysical Journal Supplement Series* **2** (1955) 1–40.

[73] Haselgrove, C. B. & Hoyle, F., "Giant stars of Type II." *Monthly Notices of the Royal Astronomical Society* **118** (1958) 519–522.

[74] "1956 July 10 meeting of the Royal Astronomical Society." *The Observatory* **76** (1956) 162–170.

[75] Cook, C. W., Fowler, W. A., Lauritsen, C. C., & Lauritsen, T., "B^{12}, C^{12}, and the Red Giants." *Physical Review* **107** (1957) 508–515.

[76] Ajzenberg, F. & Lauritsen, T., "Energy Levels of Light Nuclei. V." *Reviews of Modern Physics* **27** (1955) 77–166.

[77] Hoyle, F., "On Nuclear Reactions Occuring in Very Hot Stars. I. the Synthesis of Elements from Carbon to Nickel." *The Astrophysical Journal Supplement Series* **1** (1954) 121–146.

[78] Nakagawa, K., Ohmura, T., Takebe, H., & Obi, S., "Nuclear Reactions in the Later Stage of the Stellar Evolution." *Progress of Theoretical Physics* **16** (1956) 389–415.

[79] Salpeter, E. E., "Nuclear Reactions in Stars Without Hydrogen." *The Astrophysical Journal* **115** (1952) 326–328.

[80] Fowler, W. A., Cook, C. A., Lauritsen, C. C., Lauritsen, T., & Mozer, F., "Alpha Radioactivity of B^{12} and the Stellar Process $3He \rightarrow C^{12}$." *Bulletin of the American Physical Society* **1** (1956) 191–192.

[81] Salpeter, E. E., "Nuclear Reactions in Stars. Buildup from Helium." *Physical Review* **107** (1957) 516–525.

[82] Swann, C. P. & Metzger, F. R., "Nuclear Resonance Flourescence in O^{16}." *Bulletin of the American Physical Society* **1** (1956) 211.

[83] Bloom, S. D., Toppel, B. J., & Wilkinson, D. H., "Isotopic spin selection rules IX: The 9.58 MeV state of ^{16}O." *Philosophical Magazine* **2** (1957) 57–60.

[84] Sperduto, A., *Massachusetts Institute of Technology Laboratory for Nuclear Science Annual Progress Report* (1955).

[85] Bohr, A. & Mottelson, B. R., "Collective and individual-particle aspects of nuclear structure." *Det Kongelige Danske Videnskabernes Selskab, matematisk-fysiske Meddelelser* **27** (1953) 1–174.

[86] Gove, H. E., Litherland, A. E., & Ferguson, A. J., "The 4.97-Mev Level in Ne^{20}." *Bulletin of the American Physical Society* **3** (1958) 36–37.

[87] Hayakawa, S., Hayashi, C., Imoto, M., & Kikuchi, K., "Helium Capturing Reactions in Stars." *Progress of Theoretical Physics* **16** (1956) 507–527.

[88] Aller, L. H., *Astrophysics: Nuclear transformations, stellar interiors, and nebulæ*, volume 2, Ronald Press Co, 1954.

[89] Zanstra, H. & Weenan, J. *Bulletin Astronomical Institute of the Netherlands* **11** (1950) 165.

[90] Weenan, J. *Bulletin Astronomical Institute of the Netherlands* **11** (1950) 176.

[91] Greenstein, J. L. *Mémoires of la Société Royale des Sciences de Liège* **14** (1954) 307.

[92] Thackeray, A. D., "The helium star HD 168476." *Monthly Notices of the Royal Astronomical Society* **114** (1954) 93–100.

[93] Fowler, W. A., Burbidge, G. R., & Burbidge, E. M., "Nuclear Reactions and Element Synthesis in the Surface of Stars." *The Astrophysical Journal Supplement Series* **2** (1955) 167–194.

[94] Seitz, J. & Huber, P., "Wirkungsquerschnitt der $O^{16}(n, \alpha)C^{13}$-Reaktion für schnelle Neutronen." *Helvetica Physica Acta* **28** (1955) 227–244.

[95] Aller, L. H., *The Abundance of the Elements*, Interscience Press, 1961.

[96] Urey, H., "Boundary conditions for theories of the origin of the solar system." *Physics and Chemistry of the Earth* **2** (1957) 46–76.

[97] Urey, H. C., "The Abundances of the Elements." *Physical Review* **88** (1952) 248–252.

[98] Greenstein, J. L., "The Abundances of the Chemical Elements in the Galaxy and the Theory of Their Origin." *Publications of the Astronomical Society of the Pacific* **68** (1956) 185–203.

[99] Suess, H. E., "Über kosmische Kernbäufigkeiten. I. Mitteilung: Einige Häufigkeitsregeln und ihre Anwendung bei der Abschätzung der Häufigkeitwerte für die mittelschweren und schweren Elemente." *Zeitschrift Naturforschung Teil A* **2** (1947) 311.

[100] Suess, H. E., "Über kosmische Kernhäufigkeiten. II. Mitteilung: Einzelheiten in der Häufigkeitsverteilung der mittelschweren und schweren Kerne." *Zeitschrift Naturforschung Teil A* **2** (1947) 604.

[101] Suess, H. E., "Zur Frage nach dem Alter der Elemente." *Experientia* **5** (1949) 278–279.

[102] Suess, H. E. & Urey, H. C., "Abundances of the Elements." *Reviews of Modern Physics* **28** (1956) 53–74.

[103] Cameron, A. G. W., "Nuclear Radiation Widths." *Canadian Journal of Physics* **35** (1957) 666–671.

[104] Feshbach, H., Porter, C. E., & Weisskopf, V. F., "Model for Nuclear Reactions with Neutrons." *Physical Review* **96** (1954) 448–464.

[105] Weisskopf, V. F., "Theory of Neutron Cross Sections," in "Proceedings of the International Conference on the Peaceful Uses of Atomic Energy," volume 2 United Nations Publications, 1956 pp. 23–32.

[106] Levy, H. B., *United States Atomic Energy Commission report* **UCRL-4855** (1956).

[107] Levy, H. B., "New Empirical Equation for Atomic Masses." *Physical Review* **106** (1957) 1265–1270.

[108] Riddell, J., "A Table of Levy's Empirical Atomic Masses." *Chalk River report* **CRP-654** (1956).

[109] Cameron, A. G. W., "A Revised Semiempirical Atomic Mass Formula." *Canadian Journal of Physics* **35** (1957) 1021–1032.

[110] Cameron, A. G. W., "Nucler Level Spacings." *Canadian Journal of Physics* **36** (1958) 1040–1057.

[111] Cameron, A. G. W., in "Proceedings of the International Conference on Neutron Interactions with the Nucleus," volume CU-175 (TID-7547) Colombia University, 1957 pp. 68–71.

[112] Bidelman, W. P., "Line of the Rare-Earth Elements in the Spectrum of R Andromedæ." *The Astrophysical Journal* **117** (1953) 377–379.

[113] Buscombe, W. & Merrill, P. W., "Intensities of Atomic Absorption Lines in the Spectra of Long-Period Variable Stars." *The Astrophysical Journal* **116** (1952) 525–535.

[114] Merrill, P. W., "Spectroscopic Observations of Stars of Class S." *The Astrophysical Journal* **116** (1952) 21–26.

[115] Merrill, P. W., "Technetium in the N-Type Star 19 PISCIUM." *Publications of the Astronomical Society of the Pacific* **68** (1956) 70–71.

[116] Burbidge, E. M. & Burbidge, G. R., "Chemical Composition of the BA II Star HD 46407 and its Bearing on Element Synthesis in Stars." *The Astrophysical Journal* **126** (1957) 357–385.

[117] Hoyle, F., "The synthesis of the elements from hydrogen." *Monthly Notices of the Royal Astronomical Society* **106** (1946) 343–383.

[118] Hoyle, F., Fowler, W. A., Burbidge, G. R., & Burbidge, E. M., "Origin of the Elements in Stars." *Science* **124** (1956) 611–614.

[119] Burbidge, E. M. & Burbidge, G. R., "Relative Abundances and Atmospheric Conditions in the Magnetic Star HD 133029." *The Astrophysical Journal* **122** (1955) 396–408.

[120] Gamow, G. & Schoenberg, M., "Neutrino Theory of Stellar Collapse." *Physical Review* **59** (1941) 539–547.

[121] Baade, W., Burbidge, G. R., Hoyle, F., Burbidge, E. M., Christy, R. F., & Fowler, W. A., "Supernovæ and Californium 254." *Publications of the Astronomical Society of the Pacific* **68** (1956) 296–300.

[122] Mayall, N. U., "Nature's Greatest Explosions: Supernovæ." *The Scientific Monthly* **66** (1948) 17–24.

[123] Macklin, R. L. & Gibbons, J. H., "(p, n) Reactions in Light Nuclei I." *Bulletin of the American Physical Society* **3** (1958) 26.

[124] Swiatecki, W. J., "Systematics of Fission Asymmetry." *Physical Review* **100** (1955) 936–937.

[125] Ghiorso, A., *United States Atomic Energy Commission report* **UCRL-2912** (1956).

[126] Burbidge, G. R., Hoyle, F., Burbidge, E. M., Christy, R. F., & Fowler, W. A., "Californium-254 and Supernovæ." *Physical Review* **103** (1956) 1145–1149.

[127] Hoffleit, D., "Observation of Supernovæ." *Proceedings of the American Philosophical Society* **81** (1939) 265–276.

[128] Huizenga, J. R. & Diamond, H., "Spontaneous-Fission Half-Lives of Cf^{254} and Cm^{250}." *Physical Review* **107** (1957) 1087–1090.

[129] Borst, L. B., "Supernovæ." *Physical Review* **78** (1950) 807–808.

[130] Coryell, C. D., *Massachusetts Institute of Technology Laboratory for Nuclear Science Annual Progress Report* (1956).

[131] Parker, E. N., "The Solar Hydromagnetic Dynamo." *Proceedings of the National Academy of Science* **43** (1957) 8–14.

[132] Deutsch, A. J., "The Spectrum Variables of Type A." *Publications of the Astronomical Society of the Pacific* **68** (1956) 92–114.

[133] Burbidge, G. R. & Burbidge, E. M., "An Analysis of the Magnetic Variable α2 Canum Venaticorum." *The Astrophysical Journal Supplement Series* **1** (1955) 431–477.

[134] Meyer, P., Parker, E. N., & Simpson, J. A., "Solar Cosmic Rays of February, 1956 and Their Propagation through Interplanetary Space." *Physical Review* **104** (1956) 768–783.

[135] Craig, H., "Distribution, Production Rate, and Possible Solar Origin of Natural Tritium." *Physical Review* **105** (1957) 1125–1127.

[136] Goldberg, L., Mohler, O. C., & Müller, E. A., "The Profile of Hα during the Limb Flare of February 10, 1956." *The Astrophysical Journal* **127** (1958) 302–307.

[137] Post, R. F., "Controlled Fusion Research-An Application of the Physics of High Temperature Plasmas." *Reviews of Modern Physics* **28** (1956) 338–362.

[138] Spitzer, Jr., L., "Upper Limits on the Abundances of Interstellar Li and Be." *The Astrophysical Journal* **109** (1949) 548–550.

[139] Hunger, K. & Kron, G. E., "Polarization in a Bright-Ultraviolet T Tauri Star." *Publications of the Astronomical Society of the Pacific* **69** (1957) 347–350.